NJU SA 2021-2022

南京大学建筑与城市规划学院　建筑系教学年鉴
THE YEAR BOOK OF ARCHITECTURE DEPARTMENT TEACHING PROGRAM
SCHOOL OF ARCHITECTURE AND URBAN PLANNING NANJING UNIVERSITY
唐莲 李鑫 编　EDITORS : TANG LIAN, LI XIN
东南大学出版社・南京　SOUTHEAST UNIVERSITY PRESS, NANJING

建筑设计及其理论
Architectural Design and Theory

张 雷 教 授	Professor ZHANG Lei
冯金龙 教 授	Professor FENG Jinlong
吉国华 教 授	Professor JI Guohua
周 凌 教 授	Professor ZHOU Ling
傅 筱 教 授	Professor FU Xiao
王 铠 副研究员	Associate Researcher WANG Kai
钟华颖 副研究员	Associate Researcher ZHONG Huaying
黄华青 副研究员	Associate Researcher HUANG Huaqing
吴佳维 副研究员	Associate Researcher WU Jiawei
梁宇舒 助理研究员	Assistant Researcher LIANG Yushu

城市设计及其理论
Urban Design and Theory

丁沃沃 教 授	Professor DING Wowo
鲁安东 教 授	Professor LU Andong
华晓宁 副教授	Associate Professor HUA Xiaoning
胡友培 副教授	Associate Professor HU Youpei
窦平平 副教授	Associate Professor DOU Pingping
刘 铨 副教授	Associate Professor LIU Quan
尹 航 讲 师	Lecturer YIN Hang
唐 莲 副研究员	Associate Researcher TANG lian
尤 伟 副研究员	Associate Researcher YOU Wei

建筑历史与理论及历史建筑保护
Architectural History and Theory, Protection of Historic Buildings

赵 辰 教 授	Professor ZHAO Chen
王骏阳 教 授	Professor WANG Junyang
胡 恒 教 授	Professor HU Heng
冷 天 副教授	Associate Professor LENG Tian
史文娟 副研究员	Associate Researcher SHI Wenjuan
赵潇欣 博士后	Postdoctor ZHAO Xiaoxin
王洁琼 博士后	Postdoctor WANG Jieqiong

建筑技术科学
Building Technology Science

吴 蔚 副教授	Associate Professor WU Wei
郜 志 副教授	Associate Professor GAO Zhi
童滋雨 副教授	Associate Professor TONG Ziyu
梁卫辉 副教授	Associate Professor LIANG Weihui
施珊珊 副教授	Associate Professor SHI Shanshan
孟宪川 副研究员	Associate Researcher MENG Xianchuan
李清朋 副研究员	Associate Researcher LI Qingpeng
王力凯 博士后	Postdoctor WANG Likai

南京大学建筑与城市规划学院建筑系
Department of Architecture
School of Architecture and Urban Planning
Nanjing University
arch@nju.edu.cn   http://arch.nju.edu.cn

教学纲要（2021版）
EDUCATIONAL PROGRAM

教学阶段 Phases of Education	本科生培养（学士学位）Undergraduate Program (Bachelor Degree)			
	一年级 1st Year	二年级 2nd Year	三年级 3rd Year	四年级 4th Year
教学类型 Types of Education	通识教育 General Education			专业教育 Professional Education
课程类型 Types of Courses	通识类课程 General Courses	学科类课程 Disciplinary Courses		专业类课程 Professional Courses
主干课程 Design Courses	设计基础 Design Foundation	建筑设计基础 Architectural Design Foundation	建筑设计 Architectural Design	
理论课程 Theoretical Courses	基础理论 Basic Theory of Architecture	专业理论 Architectural Theory		
技术课程 Technological Courses				
实践课程 Practical Courses	环境认知 Environmental Cognition	古建筑测绘 Survey and Drawing of Ancient Buildings	工地实习 Practice in Construction Plant	

研究生培养（硕士学位）Graduate Program (Master Degree)			研究生培养（博士学位）
一年级 1st Year	二年级 2nd Year	三年级 3rd Year	Ph. D. Program

学术研究训练 Academic Research Training

学术研究 Academic Research

建筑设计研究 Research of Architectural Design	毕业设计或学位论文 Thesis Project or Dissertation	学位论文 Dissertation

专业核心理论 Core Theory of Architecture	专业扩展理论 Architectural Theory Extended	专业提升理论 Architectural Theory Upgraded	跨学科理论 Interdisciplinary Theory

建筑构造实验室 Building Construction Lab

建筑物理实验室 Building Physics Lab

数字建筑实验室 CAAD Lab

课程安排（2021版）
CURRICULUM OUTLINE

	本科一年级 Undergraduate Program 1st Year	本科二年级 Undergraduate Program 2nd Year	本科三年级 Undergraduate Program 3rd Year
设计课程 Design Courses	设计基础 Design Foundation	建筑设计基础 Architectural Design Foundation 建筑设计（一） Architectural Design 1 建筑设计（二） Architectural Design 2	建筑设计（三） Architectural Design 3 建筑设计（四） Architectural Design 4 建筑设计（五） Architectural Design 5 建筑设计（六） Architectural Design 6
专业理论 Architectural Theory	建成环境导论与学科前沿 Introduction to Architectural Environment and Frontiers of Disciplines	建筑导论 Introductory Guide to Architecture 城乡规划原理 Theory of Urban and Rural Planning 建筑设计基本原理 Basic Theory of Architectural Design	居住建筑设计与居住区规划原理 Theory of Housing Design and Residential Planning
建筑技术 Architectural Technology		Python程序设计 Python Programming 理论力学 Theoretical Mechanics CAAD理论与实践（一）* Theory and Practice of CAAD 1	建筑技术（一） Architectural Technology 1 建筑技术（二）声光热 Architectural Technology 2 Sound, Light and Heat 建筑技术（三）水电暖 Architectural Technology 3 Water, Electricity and Heating 可持续设计与技术* Sustainable Design and Technology
历史理论 History Theory		外国建筑史（古代） History of Western Architecture (Ancient) 中国建筑史（古代） History of Chinese Architecture (Ancient)	外国建筑史（当代） History of Western Architecture (Modern) 中国建筑史（近现代） History of Chinese Architecture (Modern)
实践课程 Practical Courses			乡村振兴建设实践 Practice of Rural Revitalization Construction 城乡认知实习 Urban and Rural Cognitive Internship
通识类课程 General Courses	数学 Mathematics 英语 English 思想政治 Ideology and Politics 科学与艺术 Science and Art 人居环境导论* Research Method of the Social Science 普通物理（力学）* General Physics (Mechanics)	社会学概论 Introduction of Sociology 城市道路交通规划与设计* Planning and Design of Urban Roads and Traffic 环境科学导论* Introduction to Environmental Science	

本科四年级 Undergraduate Program 4th Year	研究生一年级 Graduate Program 1st Year	研究生二、三年级 Graduate Program 2nd & 3rd Years
建筑设计（七） Architectural Design 7 建筑设计（八） Architectural Design 8 本科毕业设计 Graduation Project	建筑设计研究（一） Architectural Design Research 1 建筑设计研究（二） Architectural Design Research 2 研究生国际教学工作坊 Postgraduate International Design Studio	专业硕士毕业设计 Thesis Project
城市设计及其理论* Urban Design and Theory 景观规划设计及其理论* Landscape Planning and Design and Theory	建筑与规划研究方法 Research Method of Architecture and Urban Planning 城市形态与设计方法论 Urban Morphology and Design Methology 现代建筑设计基础理论 Preliminaries in Modern Architectural Design 当代景观都市实践* Contemporary Landscape Urbanism Practice 城市形态学* Urban Morphology	
建筑师业务基础知识 Introduction of Architects' Profession 建设工程项目管理 Management of Construction Project 建筑设计行业知识与创新实践* Knowledge and Innovative Practice in the Architectural Design Industry 建筑学中的技术人文主义* Technology of Humanism in Architecture CAAD理论与实践（二）* Theory and Practice of CAAD 2	建筑体系整合 Building System Integration 计算机辅助技术 Computer-Aided Design 建筑环境学与设计 Architectural Enviromental Science and Design GIS基础与应用* Concepts and Application of GIS 材料与建造* Materials and Construction 技术人文与建筑创新* Technical Humanities and Architectural Innovation* 传热学与计算流体力学基础* Fundamentals of Heat Transfer and Computational Fluid Dynamics 建设工程项目管理* Management of Construction Project 算法设计* Algorithm Design	
	建筑理论研究 Studies of Architectural Theory 建筑史方法* Method of Architectural History 中国建构（木构）文化研究* Studies in Chinese Wooden Tectonic Culture	
工地实习 Practice in Construction Plant 古建筑测绘 Survey and Drawing of Ancient Buildings		建筑设计实践 Architectural Design and Practice

课程说明：* 表示选修课程

本科建筑与规划实验班课程安排
EXPERIMENTAL UNDERGRADUATE PROGRAMME FOR ARCHITECTURE AND PLANNING

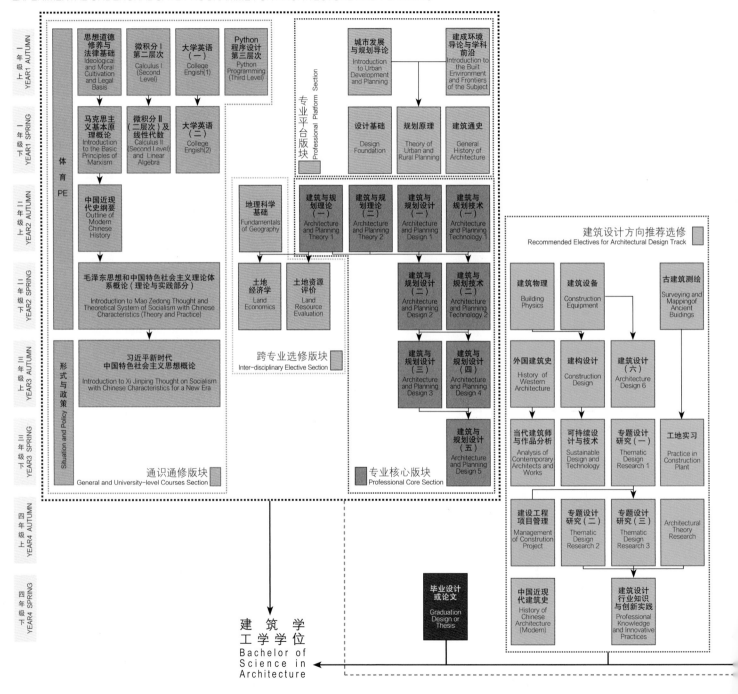

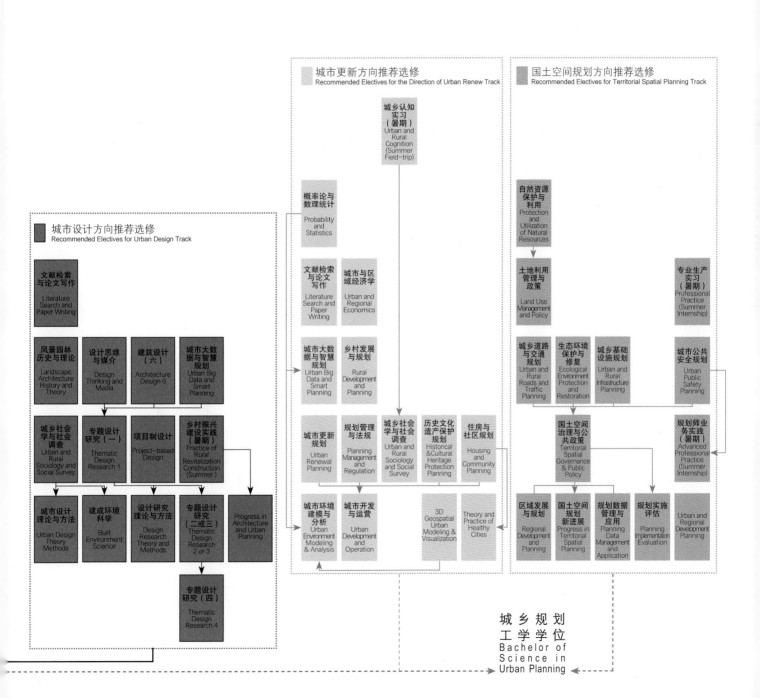

教学论文 ARTICLES ON EDUCATION

2
"空间—建造"类型学初探：与课程设计同步的建筑案例分析教学
AN EXPLORATION OF THE "SPACE-CONSTRUCTION" TYPOLOGY: ARCHITECTURAL CASE STUDIES SYNCHRONIZED WITH DESIGN TEACHING

10
高校研究生课程思政建设中的思考与探索：以"建筑体系整合"课程为例
THINKING AND EXPLORATION OF THE IDEOLOGICAL AND POLITICAL CONSTRUCTION OF GRADUATE COURSES IN UNIVERSITIES: TAKING THE COURSE OF "BUILDING SYSTEM INTEGRATION" AS AN EXAMPLE

课程概览 COURSE OVERVIEW

24
建成环境导论与学科前沿
INTRODUCTION TO ARCHITECTURAL ENVIRONMENT AND FRONTIERS OF DISCIPLINES

26
设计基础
DESIGN FOUNDATION

36
建筑设计基础
ARCHITECTURAL DESIGN FOUNDATION

40
建筑设计（一）：限定与尺度：独立居住空间设计
ARCHITECTURAL DESIGN 1: LIMITATIONS AND SCALES: INDEPENDENT LIVING SPACE DESIGN

44
建筑设计（二）：校园多功能快递中心设计
ARCHITECTURAL DESIGN 2: CAMPUS MULTIFUNCTIONAL EXPRESS CENTER DESIGN

48
建筑设计（三）：专家公寓设计
ARCHITECTURAL DESIGN 3: THE EXPERT APARTMENT DESIGN

52
建筑设计（四）：世界文学客厅
ARCHITECTURAL DESIGN 4: WORLD LITERATURE LIVING ROOM

56
建筑设计（五）：大学生健身中心改扩建设计
ARCHITECTURAL DESIGN 5: RECONSTRUCTION AND EXPANSION DESIGN OF THE COLLEGE STUDENT FITNESS CENTER

64
建筑设计（六）：社区文化艺术中心设计
ARCHITECTURAL DESIGN 6: DESIGN OF COMMUNITY CULTURE AND ART CENTER

72
实践课程：工地实习
PRACTICAL COURSES: PRACTICE IN CONSTRUCTION PLANT

78
建筑设计（七）：高层办公楼设计
ARCHITECTURAL DESIGN 7: DESIGN OF HIGH-RISE OFFICE BUILDINGS

84
建筑设计（八）：城市设计
ARCHITECTURAL DESIGN 8: URBAN DESIGN

90
本科毕业设计
GRADUATION PROJECT

102
建筑设计研究（一）：基本设计
ARCHITECTURAL DESIGN RESEARCH 1: BASIC DESIGN

120
建筑设计研究（一）：概念设计
ARCHITECTURAL DESIGN RESEARCH 1: CONCEPTUAL DESIGN

132
建筑设计研究（二）：综合设计
ARCHITECTURAL DESIGN RESEARCH 2: COMPREHENSIVE DESIGN

144
建筑设计研究（二）：城市设计
ARCHITECTURAL DESIGN RESEARCH 2 : URBAN DESIGN

156
研究生国际教学工作坊
POSTGRADUATE INTERNATIONAL DESIGN STUDIO

目　录

1—21 教学论文　ARTICLES ON EDUCATION

23—165 课程概览　COURSE OVERVIEW

167—178 建筑设计课程　ARCHITECTURAL DESIGN COURSES

179—182 建筑理论课程　ARCHITECTURAL THEORY COURSES

183—186 城市理论课程　URBAN THEORY COURSES

187—190 历史理论课程　HISTORY THEORY COURSES

191—196 建筑技术课程　ARCHITECTURAL TECHNOLOGY COURSES

197—199 认识实习　COGNITIVE INTERNSHIP

201—207 其他　MISCELLANEA

教学论文
ARTICLES ON EDUCATION

教学论文 ARTICLES ON EDUCATION

"空间—建造"类型学初探：与课程设计同步的建筑案例分析教学
AN EXPLORATION OF THE "SPACE-CONSTRUCTION" TYPOLOGY: ARCHITECTURAL CASE STUDIES SYNCHRONIZED WITH DESIGN TEACHING

吴佳维　冷天　孟宪川　麦思琪
WU Jiawei, LENG Tian, MENG Xianchuan, MAI Siqi

摘 要

对结构、构造知识的追寻应建立在主动的设计意识基础之上，设计教学对建造议题的讨论不应受知识掌握程度的绝对制约。本文提出了一种从探寻"空间—建造"关系出发、以案例分析为手段、以关键词为抓手、与设计过程深度结合的设计教学方法。

Abstract

The pursuit of structure and construction knowledge should be based on active design awareness. The discussion of construction issues in design teaching should not be confined by the students' level of construction knowledge. This paper proposes a design pedagogy that applies the architectural case study to explore the "space-construction" relationship, while using keywords as a tool, and deeply integrated with the students' design process.

图1 三种"空间—建造"类型

一般而言，设计教学会通过大跨度、大空间来讨论"空间—结构"议题，因而对学生在构造、结构方面知识的掌握要求较高，相应的课程设计往往在三年级以上展开。其实，对结构、构造知识的追寻应建立在主动的设计意识上，那么，如何在低年级设计教学中建立学生的"空间—结构—材料"意识，促进他们未来对建构议题的求索呢？南京大学建筑学本科的构造课程设在三年级第一学期，但大二学生着手的第一个课程设计已经要求在大比例的剖透视图中表达构造关系[1]，这对学生和教师都是一个挑战。一种策略是给定建筑的结构体系及外墙做法，使学生专注于适应于场地和使用需求的空间组织，后期再结合构造做法调整设计，近年来"建筑设计一：小住宅设计"采取这种方法使得初学者也能够完成较有深度的设计。在此基础之上，教学小组在本年度"建筑设计二：校园快递中心设计"中进行了新的尝试。

1. 原理：建造方式的三种原型及其空间特征

在德语系国家有一种常用的构造方式分类法，按照承重构件的形态和布置特征，将那些以体块式构件堆砌或浇筑而成的墙体承重并构成封闭空间、沉重体量的建造方式归类为实体式建造（德：Massivbau；英：solid construction）；将那些由纤细的杆件或线性元素编织而成的建造方式称为杆系式建造（德：Filigranbau；英：skeleton construction），其构成的空间往往较为轻盈通透；还有一种实体式建造的特殊状态，即承重墙只出现在单个走向上的平行板式建造（德：Schottenbau；英：slab construction）[2]。有趣的地方在于，这种分类方法并不由承重的材料所决定，而明显地与其表现出来的空间特征相关。例如，一般认为，木材是一种杆件形态的材料，因此木构建筑所当然应属杆系建造，然而，并干式木构以砌筑的方式堆叠木材，且形成了四角封闭的包裹式空间，故被归类为实体式建造。不但如此，对于蒙古包的构造方式，一些学者因其外墙的支撑结构是由纤细材料编织而成而认为其应属杆系式建造，而另一些学者根据其复合的包裹结构及密闭性将其归类为实体式建造[3]。可见，这种建造方式的分类包含了人的感知【图1】。

上述三种建造方式分别构成了包裹式（内向性）空间、流动式（多向性）、引导式（定向性）空间。在这三个原型中，承重元素也是主要的空间限定元素，也就是说，结构形式决定了空间特质。除此之外，三个原型之间存在着过渡的形式。笔者认为，正是在三个原型及其变体与具体建筑材料的结合之中蕴含了丰富的形式可能。在教学中，这些"空间—建造"的基本关系能够帮助学生识别当代建筑设计形式中的基本规律，进而使他们摆脱"功能—形式"的单一赋形途径以及肤浅的形式模仿。

2. 案例分析结合课程设计

作为教学法的一种，与设计进度并行的案例分析能够使学生带着设计问题去理解案例中建筑师的具体设计策略，寻找可能的解答。要达到这样的效果，要求案例选择与课程设计任务具有可类比性。这一相似性不必是建筑类型上的，而重在建筑规模及空间组织关系。

本次设计题目是约200 m^2 的"校园快递中心"，除了具有与学生的日常生活关系紧密、功能流线简明的特点外，该题目还包含2个隐含信息：其一，快递柜或快递架尺寸具有规律性，能为结构布置提供模数参照；其二，快递柜本身有成为空间限定元素的潜质，而不是一种随意摆放的家具。此外，设计任务还在快递收发、员工休息区之外，加入了校园复合空间（面积及容纳的活动内容由学生自定）。

Ex. 1 案例精读　阶段一
结构布置 Structural layout

主结构　　　　　　　　　　Primary structure

空间域　　　　　　　　　　Spatial zoning
运动趋势　　　　　　　　　Movement tendency

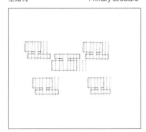
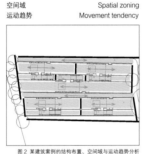

图 2　某建筑案例的结构布置、空间域与运动趋势分析

课程设计任务书

第一周　场地认知，结合已有图像资料对快递服务中心的建设场地进行实地调研。
第二周　制作场地模型，比例 1:100。提出设计概念。
第三周　思考与场地周围现有建筑的对话关系，提出处理设计问题的结构策略，整合结构与空间的组织方式。建立初步方案与工作模型。
第四周　深化方案阶段，优化并发展前述的结构策略，用 1:50 的图纸比例，手绘平立剖面图纸，在初步方案的基础上深化实体与建造层面的思维。
第五周　建立工作模型辅助设计，进一步优化结构设计，使得结构部分清晰明确可认知，明确承重结构和围护的各自作用。
第六周　确定最终的设计方案，并将研究的重心转移到建构设计研究部分。
第七周　了解实际建造所面临的误差问题和节点问题，推敲各设计细节，制作全新的结构体模型，比例 1:50。
第八周　排版调整，思考并选择图面表达的效果，制作必要的分析图和效果图。整理并完成图纸，制作正式模型（基地模型 1:100）并完成课程答辩。

案例精读时间表

Ex. 1 案例精读 阶段一	Lect. 1 空间限定
	Lect. 2 结构与空间
Ex. 2 案例精读 阶段二	Lect. 3 材料与营造
	Lect. 4 细部读图

图 3　案例分析与课程设计进度的关系

从建筑类型上来说，目前少有小型快递站建筑范例，但类似尺度、相似的"服务—被服务"空间组织关系却可以在当代众多画廊、独立住宅等小型建筑中找到类比。我们筛选出 9 个国内外的建筑案例，它们从建造方式上涵盖了前文所述的实体式建造和杆件式建造类型，从建筑材料上包括了砖石结构、木结构、钢结构，从项目环境上包括了热带、温带、地中海气候等多个类型，在技术上有应用了精细构造做法获得抽象效果的，也有巧妙运用乡土建造技术的地域性表达。

每一个案例由一组同学（2~3 人）完成，分为前后 2 个阶段。第一阶段对应于课程设计的概念构思，分析场地操作和空间组织；第二阶段对应课程设计深化，分析案例的材料选择、结构体系和构造层级，以及它对设计概念的呼应，如此完成剥洋葱式的进阶理解。

3. "关键词 + 图解"引导概念理解

在过往教学中发现，仅仅规定分析的议题并不足以帮助学生剥离出准确的信息点。低年级学生对建筑专用术语陌生，无法准确描述自己或他人的设计，更无法用这些概念指导设计。因此，我们在本次教学中采取关键词与分析图解双重限定的模式，要求学生按规定方式绘制分析图解来表达特定关键词。例如，在"场地操作"分析中，以黑色块表达建筑实体，以留白表达场地，呈现"图—底"关系；在"空间组织"分析中，以灰色块和红色块分别表示"被服务空间"与"服务空间"，体现平面布置对"使用需求"的回应方式。

在第一阶段分析中比较特别的是，将"空间域"（spatial zoning）与"运动趋势"（movement tendency）两个关键概念置于"结构布置"分析议题下。这两个概念均来自得克萨斯州骑警核心成员伯纳德·霍斯利（Bernhard Hoesli）于 1960 年代开始在苏黎世联邦理工学院建筑系进行设计入门教学。在他给学生列出的《基本概念》名词释义中，"空间域是空间的局部，在空间内部发挥空间效果。空间定义元素的作用令空间域的界限可被感知"，而"运动趋势是一种在地的空间特征，它通过空间构成元素的布局及大小比例引导观者朝大致某个方向移动"[4]。在霍斯利的概念中，空间定义元素与其是否起承重作用无关；在本次案例分析中，要求学生识别出承重构件，去除非承重的元素，在简化的平面上将柱或承重墙填红色，并用点画线标出梁的轴线位置。承重构件的疏密程度划分了"空间域"，同时构成了空间的方向性【图 2】。学生通过读图、分析，能够发现不同的建造方式在结构布置上的特点和相应的空间格局。由此作为分析第一阶段的结束。

分析的第二阶段，首先要求学生详细阅读案例的构图，标注关键的构造信息；比照建筑照片提取最主要的建筑材料，标出相应的构造节点，绘制承重结构的轴测图；根据功能性区分构造层，以颜色区分保温隔热层、内饰面、保护层，以传达"功能层应连续"的构造基本原理；最后，根据前述分析，呈现案例局部的"构造—结构—空间"关系。在此阶段，不同建造方式所带来的建筑结构与气候边界之重合或分离的关系、流动空间与单元式空间之区别等特征凸显出来。

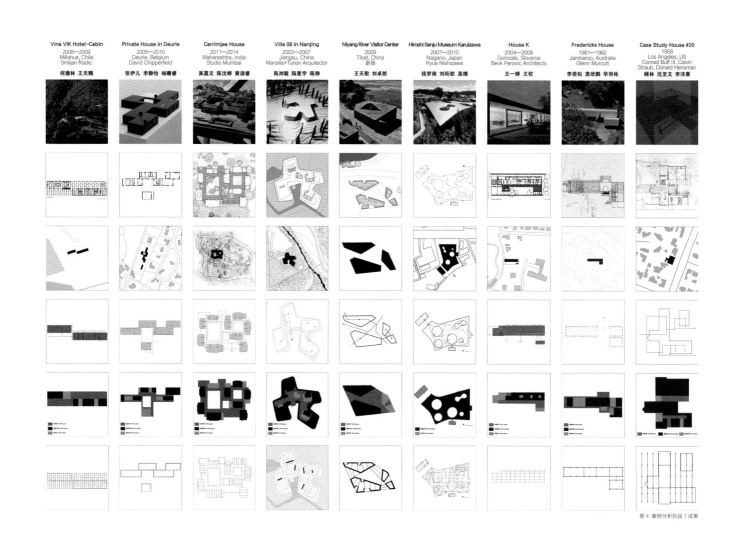

图4 案例分析阶段1成果

案例分析与课程设计题目同时布置,并辅以专题讲座。在总共8周的课程时间内,学生在第2~3周的课堂上展示第一阶段的分析,每个小组约10 min;在第5~6周展示第二阶段的解读【图3】。如此,整个班级构成了学习的共同体,学生对"空间—建造"类型的特征和构造基本关系有了总体的认识【图4、图5】。一些学生能够将这种空间认识运用到自己的设计中,方案体现了结构体系与空间的整合关系,还有一部分学生能够在结构布置的规律中发现形式的创新点,并结合合理的构造做法获得良好的空间品质【图6】。

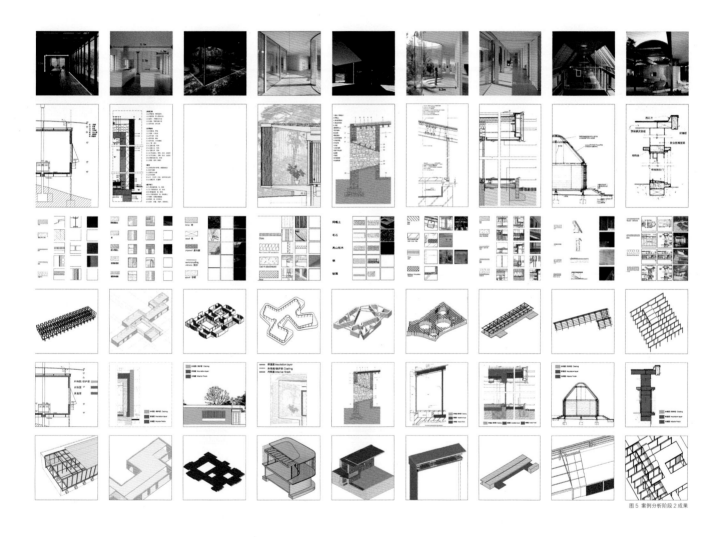

图 5 案例分析阶段 2 成果

4. 总结与反思

  本教学小组探寻了一种使当代建筑设计变得"可教"的途径。案例分析的过程试图让学生紧扣"空间—建造"议题，专注于读图、识图，在理解之后，以最小的图量做表达，避免学生在埋首绘制一个又一个分析图的过程中实质上一步步远离了设计初衷。本学年是该教案的第一次实施，取得的阶段性效果是：学生作为初学者能够较为准确地掌握与设计相关的关键概念，较快进入了建筑设计的语境；对构造产生了兴趣，理解了材料、建造与设计的紧密关系。种种不足在实施过程中展现出来，例如案例基本图纸资料的部分缺失导致低年级学生无法重现构造做法；由于构造基本知识的缺乏，阅读技术图纸困难；更主要的是，能够在案例分析与自己的设计之间建立有效关联的学生属于少数。未来的教学将以读图辅导、实体模型制作等方式针对性地补足这些方面并应对新的课程体系做调整。

In general, as the "space-construction" issue is usually discussed in a large span and a large space in design teaching, students are highly required to master the knowledge of construction and structure, and the corresponding course design is often started from the third year. Actually, the structure and construction knowledge should be pursued based on active design awareness. Then, how should we build students' awareness of "space-construction-material" in the design teaching of lower grades to promote their future exploration of the construction issue? The undergraduate construction course is set in the first semester of the third year in the architecture major of Nanjing University, but sophomores have been required to express the construction relations in a large-scale profile view in their first course design of sophomores[1], which is a challenge for both students and teachers. One strategy is to give the structural system of the building and the method of making the outer wall, so that students can focus on the spatial organization adapted to the site and usage demand, and then adjust their design according to the construction method later. In recent years, this method has been adopted in Architectural Design 1: Design of Small Houses so that beginners can also complete in-depth design. On this basis, the teaching team made a new attempt in Architectural Design 2: Design of Campus Express Center this year.

1. Principle: three prototypes of construction modes and their spatial features

In German-speaking countries, there is a common construction classification method, where according to the forms and layout of load-bearing members, the method in which load-bearing walls are stacked or poured by block-type members and form a closed space and heavy volume is classified as solid construction (German: Massivbau); the method in which the construction is woven by thin rods or linear elements is called skeleton construction (German: Filigranbau), and the space is often transparent; there is also a special state of physical construction, namely slab construction (German: Schottenbau) in which the load-bearing walls only appear on a single direction[2]. Interestingly, this classification method is not decided by the bearing materials, but is obviously related to the spatial features. For example, timber is generally considered a material in the form of rods, and thus wooden architecture should be skeleton construction. However, the log cabin timber framework is stacked by timbers in a masonry manner, and forms a wrapped space with four corners closed, so it is classified as solid construction. Moreover, some scholars classify yurts as skeleton construction since the supporting structure of their outer walls is woven by thin materials, while some others, based on their composite wrapped structure and leakproofness, classify them as solid construction[3]. This shows that this classification method is subject to human perception [Figure 1].

These construction methods constitute wrapped (inward directional) space, flowing space (multi-directional) and guiding (directional) space. Among the three models, the load-bearing element is also the main space confinement element, that is, the structural form determines the spatial features. In addition, there are also transitional forms among the three models. The author believes that there may be abundant forms in the combination of the three models and their variants and the specific building materials. In teaching, these basic "space-construction" relations can help students identify the fundamental rules of modern architectural design forms, and thus avoid their single forming approach of "function-form" and superficial form imitation.

2. Case analysis combined with curriculum design

As one of teaching methods, case analysis conducted with the design process can enable students to understand the specific design strategies of the architect in the case and find possible answers with design questions. To this end, the selected case should be comparable to the design task of the course. This similarity is not necessarily in the architectural types, but the architectural scale and spatial organization relations.

The title of this design is "Campus Express Center" of about 200 square meters. In addition to the center's close relation to students' daily life and its simple function zones, this title also implies two pieces of other information: first, the express cabinet or express rack is of regular size, and can provide modular reference for structural layout; second, the express cabinet is a potential space confinement element, rather than a piece of randomly placed furniture. Besides, the campus composite space (the area and the activities to be conducted are determined by students) is also added in the design task in addition to package receiving and sending area and staff rest area.

In terms of architectural types, there are few examples of small express stations, but analogies can be found in modern galleries, separate houses and other small buildings of similar scale and similar "service-serviced" spatial organization relation. We selected 9 domestic and foreign architectural cases, which covered the solid construction and skeleton construction types described above in construction methods, masonry, timber and steel structures in construction materials, and tropical, temperate and Mediterranean climates in project environments. Technologically, some buildings achieve an abstract effect by fine construction method, and some are regionally expressed by local construction technique.

Each case was completed by 2-3 students in a group, and was divided into two stages. The first stage, corresponding to the conceptual design in course design, analyzed the site operation and spatial organization; the second stage, corresponding to the course design development, analyzed the material selection, structural system and construction levels of the case, and the correspondence to the design concept. In this way, students could understand relevant knowledge more deeply.

3. Conceptual understanding under the guidance of "keywords + diagrams"

Findings in past teaching showed that it was not sufficient to help students to find accurate information points by only specifying the topic to be analyzed. Lower-grade students unfamiliar with architectural terminology were unable to accurately describe their own or others' design, or guide design with these concepts. Therefore, we defined keywords and diagrams simultaneously in this teaching, and asked students to draw diagrams in a prescribed way to express specific keywords. For example, in the analysis of "site operation", the black block was used to express the building entity, and the white space was used to express the site, to represent the "diagram-bottom" relation; in the analysis of "spatial organization", the grey block and red block were used to represent the "serviced space" and "service space", to reflect the correspondence of the plan layout to "usage demand".

Specially, in the first stage, the two key concepts of "spatial zoning" and "movement tendency" were placed under the topic of "structural layout". They were from the introduction to design taught by Bernhard Hoesli, the core member of Texas Ranger Division, in Eidgenö ssische Technische Hochschule Zürich since the 1960s. In the glossary of *Basic Concept* he listed for students, "Spatial zoning is a part of space, and plays a spatial effect inside the space. The space definition element makes the boundaries of spatial zoning perceptible." "Movement tendency is a spatial feature on the ground, which guides the viewers to move in a general direction through the layout and size of spatial elements." [4] In Hoesli's concepts, the space definition element is independent of its role in bearing loads; in this case analysis, students were required to identify load-bearing members, eliminate non-load-bearing elements, paint the column or load-bearing walls red on a simplified plan, and mark the axis of the beam with dot lines. The density of load-bearing members divided the "spatial zoning", and also constituted the directivity of the space [Figure 2]. Students could find the features of different construction methods in the structural layout and the corresponding spatial pattern by interpreting diagrams and analysis. The first stage was completed.

In the second stage, students were asked to interpret the construction diagram of the case in detail and mark the key construction information; then they were asked to extract the most important architectural materials based on the architectural figures and mark the corresponding construction nodes; they should make the axonometric drawing of the load-bearing structure; they should divide the construction layers according to functionality and mark the thermal insulation layer, interior layer and protection layer in different colors, to convey the basic construction principle that "the functional layers should be continuous"; finally, students were asked to represent the part "construction-structure-space" relation of the case. In this stage, features such as the overlapping or separation between the architectural structure and the climate boundary brought by different construction methods, and the differences between the flowing space and the unit space were highlighted.

Case analysis and course design subjects were assigned simultaneously, complemented by thematic lectures. In the 8 weeks of course, students presented their analysis of the first stage in class in week 2-3, with abuot 10 minutes for each group; then they presented the interpretation of the second stage in class in week 5-6 [Figure 3]. In this way, the whole class became a learning community, where students had a general understanding of the features and basic construction relations of "space-construction" [Figures 4 and 5]. Some students could apply their spatial understanding to their own design and reflected the integration between the structural system and space in their plan, and some could find the innovation points of forms in the rules of the structural layout, and obtain good space quality by reasonable construction methods [Figure 6].

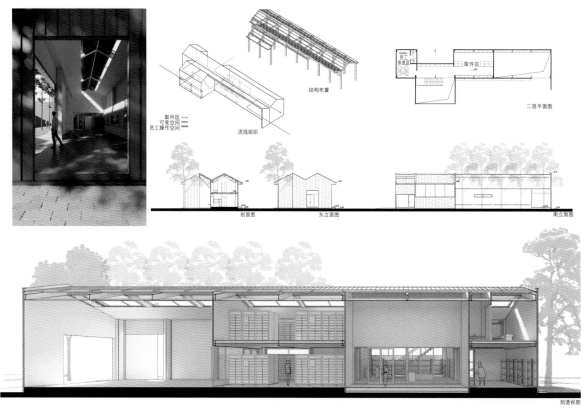

图 6 学生作业（学生姓名：沈至文 2020 级建筑）

4. Summary and reflection

This teaching team explored an approach to make the modern architecture design "teachable". The case analysis intended to make students keep the topic of "space-construction" and focus on diagram interpretation and identification. After students, understanding, the least number of drawings was used for expression to prevent students from making several diagnosis diagrams and straying from the original design intention. This plan was conducted for the first time in this academic year, and the results achieved were as follows: as beginners, students could accurately master key design-related concepts, and get into the context of architectural design quickly; they became interested in construction, and understood the close relations between materials, construction and space. There were also deficiencies in the implementation. For example, lower-grade students were unable to reproduce the construction method due to the missing of some basic drawing data of the case; students found it difficult to interpret technical drawings for lack of basic construction knowledge; more importantly, few students could establish an effective association between the case analysis and their own design. In future teaching, we will adjust the plan to teach students to interpret drawings and make solid models to make up for these deficiencies in a targeted way and correspond to the new course system.

参考文献

[1] 至 2021/2022 年度，南京大学建筑学本科课程设计开始于二年级下学期。一年级为设计通识课程，二年级上学期为不含建筑设计内容的测绘、分析作业。

[2] 德文与英文称法并不完全对应，此处不细述。

[3] MEIJS M, KNAACK U. Components and connections: Principles of construction[M]. Basel: Birkheuser, 2009.

[4] 翻译自霍斯利的《基本概念》，档案来自苏黎世联邦理工学院建筑系建筑历史与理论组的 GTA Archive。

教学论文 ARTICLES ON EDUCATION

高校研究生课程思政建设中的思考与探索：
以"建筑体系整合"课程为例

THINKING AND EXPLORATION OF THE IDEOLOGICAL AND POLITICAL CONSTRUCTION OF GRADUATE COURSES IN UNIVERSITIES: TAKING THE COURSE OF "BUILDING SYSTEM INTEGRATION" AS AN EXAMPLE

吴蔚
WU Wei

摘 要

目前我国的课程思政建设主要集中在本科教育阶段，研究生思想政治教育才刚刚起步，但作为更高层次的高等教育，研究生思想政治教育也是高校课程思政教育中的重要一环。论文以南京大学建筑与城市规划学院所开设的研究生建筑技术课程"建筑体系整合"为例，探讨如何根据课程本身的特点和研究生的精神特质，从课程定位、教学内容、教学模式、考核方式四个方面开展课程思政教学改革。课程思政建设的关键和起点是课程定位，"建筑体系整合"课程首先围绕着"立德树人"这一教育主线被重新定位，教学体系和教学内容也随之进行重新建构和整合，其主要目标是打破建筑技术专业理论与建筑师价值观和人生观培养之间的壁垒，打破专业知识传授与思想教育培养之间的壁垒。在教学模式上则根据学生的需要和兴趣，以培养建筑师职业修养和正确价值观为目标，是一种开放式、双向式的灵活教学模式。此外，考核和考评模式也是配合课程思政教育的重要手段。教改实践显示只有在重新建构课程体系的基础上，结合学生的实际需求，从多方位、在多层次上有机地融入思政元素，通过潜移默化地引导学生建立正确的人生观和价值观，才能较好地将立德树人的精神贯穿在高等专业课程教育之中。

Abstract

At the present, construction of curriculum-based ideological and political education is mainly focused on the undergraduate teaching, and it has just started for postgraduate study. As an advanced education in universities, its ideological and political study is also important. Take a graduate course "Building System Integration" in the school of architecture and urban planning of Nanjing University as an example, the paper discusses how to carry out the reform of ideological and political teaching. According to the course's nature and spiritual characteristics of the graduate students, the reform is conducted from four aspects: the course orientation, teaching content, teaching mode and assessment method. The key and the starting point are to reposition the course. The course is first repositioned around the theme of "Strengthening Morality Education". And the teaching content has also been reconstructed and reintegrated based on the new teaching theme. Its main purposes are to break the barriers between building technology theory and the cultivation of architects' outlook on values and life, as well as between professional knowledge and ideological education. According to the needs and interests of students, it attempts to develop an open and two-way flexible teaching method which aims to cultivate architects' professionalism and correct values. In addition, the assessment and evaluation mode is a useful tool to cooperate with the reform. This study shows that a successful reform should be based on reconstructing the curriculum system. According the actual needs of postgraduate students, it is necessary to integrate ideological and political elements naturally into multiple directions and levels of a course. Be imperceptibly guiding students to establish a correct outlook on life and values, it is better to integrate the spirit of establishing morality and cultivating people into higher professional curriculum education.

为深入贯彻落实习近平总书记关于教育的重要论述和全国教育大会精神，以及中共中央办公厅、国务院办公厅《关于深化新时代学校思想政治理论课改革创新的若干意见》、教育部《高等学校课程思政建设指导纲要》文件中的精神，我国各个大学在近年来都开始全面推动课程思政建设[1-3]。建筑学是一个跨学科的专业，面临着一系列综合、复杂的技术问题和社会问题。对于新一代的建筑师，不仅仅需要对其加强职业教育，也必须跟上时代的脚步培养其正确的价值观和职业修养。正如邱开金所指出的"课程思政建设可以帮助养成正确的人生观、价值观"[4]，而只有建立正确的人生观和价值观，才能培养出德才兼备的合格的建筑师。

目前我国的课程思政教学改革主要集中在本科教育阶段，研究生思想政治教育才刚刚起步。尽管研究生教育仍属职业教育范畴，但和本科教育相比仍有很大的区别[5]。研究生教育强调对事物认知能力的培养，研究生的心理和生理年龄较本科生成熟，因此简单地套用本科课程思政建设的经验会出现很多问题。既往研究显示，研究生的课程思政教改存在着教学方式显性化、形式化与功利化的困境[6]，因此有必要结合研究生培养目标和培养方式，根据研究生群体的精神特质和价值追求，探索一条有效开展研究生课程思政建设的发展思路。本文以南京大学研究生院积极推动的研究生课程思政教学改革为契机，以笔者所教授的建筑技术研究生课程"建筑体系整合"为例，尝试从课程定位、教学内容、教学模式、考核方式四个方面探讨如何将课程思政建设有机融入研究生教学中，在分享思政教改中的经验和教训的基础上，探索更有操作性和实效性的研究生课程思政教育。

1. "建筑体系整合"课程思政建设所面临的困难

"建筑体系整合"是南京大学建筑与城市规划学院所开设的研究生课程，其主要内容涵盖了建筑体系整合理论和建筑使用后整体评价，以及建筑技术理论研究方法三部分知识内容。该课程的主要教学目标是通过学习建筑技术理论知识和使用后评价方法，让学生采用整体建筑评价方法发现和分析既有建筑所存在的实际问题，从而引导同学们重新审视和评判建筑设计的理念和价值系统，树立建筑环境意识和可持续发展价值观。尽管该课程为南京大学建筑与城市规划学院建筑技术方向研究生的必修课程，但也对南京大学所有研究生开放选修。

尽管"建筑体系整合"课程在总体教学目标上与中共中央办公厅、国务院办公厅《关于深化新时代学校思想政治理论课改革创新的若干意见》的精神相吻合，但作为一个专业性较强的工科课程，在实施课程思政教改上面临着以下三个方面的困难：

首先，在教学内容上"建筑体系整合"是一门新课程。该课程借鉴了新加坡国立大学、香港中文大学、美国康奈尔大学和麻省理工学院等西方高校的先进教学内容。这些教学内容涵盖了大量国外较前沿的建筑专业理论知识，不仅专业性强，也是一门较新的建筑专业课程，目前也只有南京大学建筑学系开设此课程，其主要专业教材、讲义和阅读材料都是英文，只有少部分教学内容被翻译成中文，因此在挖掘课程思政元素上有一定难度。

其次采用了较新的教学模式。该课程不仅仅是建筑技术专业研究生的必修课程，也是一门供建筑与城市规划学院，甚至是全校研究生开放选修的研究生课，选修"建筑体系整合"课程的研究生来自南京大学各个院系，除了一些传统的工科院系外，有些理科和文科专业如数学、经管类专业的研究生也来选修此课程，学生背景复杂，知识水平参差不齐，为配合来自不同背景的研究生，许多较新的教学模式被采用，如结合慕课的线上线下混合教学、翻转课堂等。但如何将课程思政教学应用在新型教学模式上，目前只有较少的研究[7]。

最后是教学时间和教学安排。南京大学建筑与城市规划学院所开设的研究生课程较多，因此在教学安排上非常紧凑，如"建筑体系整合"课程被安排在学期的前8周，每次课堂时间为4个小时。课程内容多、课程安排紧凑且每次上课时间长，这对于学生和老师都是一个挑战，特别是外院系选修课程的学生。

尽管"建筑体系整合"在课程内容还是教学模式上，均借鉴了西方先进国家的培养模式，但无论如何改变，其课程定位、教学内容、教学模式和考核方式这四个课程基本要素主导了课程的整体建构。因此，要想开展"建筑体系整合"的课程思政建设，也需要从这四个方面展开。

2. 在课程定位上强调课程思政内容

所谓"课程思政"建设，就是从大思政格局中出发的一场教育改革的实践，是将立德树人这一教育目标贯穿于课程体系的各个环节，通过激发课程中的思政元素，将知识传授与价值引领相结合，引导学生向德、智、体、美、劳全面发展[8]。课程思政建设的关键和起点是课程定位，要想让研究生专业教育与思政教育相互融合，必须重新定位该课程的培养目标和教学内容，重新建构课程体系，拓展新的教育内涵，才能让学生学习专业知识的同时，建立正确的价值观和世界观。

建筑专业教育培养的是未来的建筑师，但我国建筑教育界目前存在着不少问题，如以建筑设计为主导的专业教学模式，往往造成建筑学学生重艺术、轻技术，重形态分析、轻客观量化，理论与实际脱节，这导致培养出来的建筑师思想不够坚定，对我国国情、社情了解不够，因此存在着"贪大、媚洋、奇怪"等诸多建筑乱象。早在2014年10月15日，中共中央总书记、国家主席、中央军委主席习近平在北京主持召开文艺工作座谈会时表示，就指出不要搞"奇奇怪怪的建筑"。

针对这些问题，"建筑体系整合"的课程思政建设的定位是从两个打破开始，首先是打破建筑技术专业理论与建筑师价值观和人生观培养之间的壁垒，其次是打破专业知识传授与思想教育培养之间的壁垒，尝试将立德树人的精神贯穿在专业课程教育之中，引导学生将理论化为实践，将知识践于德行，成为一位真正的社会主义建设者。通过打破这两个壁垒，笔者对"建筑体系整合"课程重新予以定位，即在借鉴西方先进知识的前提下，落实我国"适用、经济、绿色、美观"的新时期建筑方针，同时坚定文化自信，增强文化自觉，延续中华文化文脉，在展现时代精神的同时，也要学习体现地域文化特色。同时，课程也要从各个方面加强学生的思想教育，特别是激发建筑学学生的科学精神，培养学生的职业素养。

3. 重新建构和整合"建筑体系整合"教学内容

"建筑体系整合"专业理论知识包括建筑技术基础理论、建筑体系整合理论、建筑使用后整体评价三部分内容，其中建筑技术基础理论知识为笔者所开设的慕课，采用线上教学，不易改动。线下教学首先讲解的是建筑体系整合理论，即什么是建筑体系整合，以及如何从整合的角度解读各个建筑体系，和这些体系之间相互影响、相互制约的关系。课程其次介绍的是建筑使用后整体评价方面的理论和方法，即利用建筑整合的理念来分析和评估建筑整体环境。

根据新的课程定位，笔者针对线下教学的主要内容上进行重新建构和整合，新旧课程内容和所占课时比较详见表1，课程内容围绕着对学生的价值引领和品德培养。如在课程开始讲授建筑技术发展史时，既往只是讲工业革命后西方建筑发展史，而现在则加入我国五千年历史长河中所涌现的优秀古建筑和先进技术，特别是我国民居是如何在选址、空间布局、建造技术等许多方面遵循着建筑与环境、建筑与人、建筑与社会的和谐关系。

为了让"建筑体系整合"这门课程真正实现本土化，在讲授建筑体系整合基础理论时，有意识地结合着我国最新的国情、社情进行介绍，从最新的中共中央办公厅、国务院办公厅印发的《关于推动城乡建设绿色发展的意见》到我国"十四五"发展战略等，让学生学习生态优先/绿色发展的理念。此外，在让学生学习建筑使用后整体评价理论的同时，也提出用生态和理性的眼光，重新审视我国建筑的发展现状，对比我国传统民居，在处理人与自然、人与人、人与资源关系的巧妙之处，挖掘渗透在其中朴素的生态建筑思想和蕴含在其中的人文意识及和谐理念，进而促进、完善和发展现代建筑文化，支持并形成健康、科学的现代建筑文化理念及实践。

表1 课程思政教改前后教学内容与课时对比

原教学内容	思政课程教改后的教学内容
建筑体系中的美和艺术：工业革命后西方建筑技术发展史	增加我国五千年历史长河中所涌现的优秀古建筑和古人在建造技术上的辉煌成就，以及今天我国建筑智能化水平的蓬勃发展现状
课时：4个课时	课时：4个课时
建筑体系整合理论	在原有理论知识基础上，增加我国现行的各种建筑规范，以及我国建筑的发展趋势，从最新的中共中央办公厅、国务院办公厅印发的《关于推动城乡建设绿色发展的意见》到我国"十四五"发展战略等，从这些最新的国情、社情介绍，让学生学习生态优先、绿色发展的理念
课时：10课时	课时：10课时
建筑使用后整体评价理论	让学生学习从建筑使用者的角度、以人为本地认识建筑，让学生更多地从建筑的实际使用出发，从整体建筑环境、可持续发展方面去考虑规划和建筑设计，而不仅仅是从空间使用、艺术造型来进行建筑设计
课时：12课时	课时：12课时
实践认知	实践认知从原有的2个学时增加到6个学时，充分利用南京本地的红色文化资源，将学生带到红色教育基地，在红色教育基地，进行了室内空间、声、光、热、室内空气质量等一系列测量，进行建筑使用后整体评价
课时：2课时	课时：6课时
学生汇报	1）将个人作业改成小组作业，培养学生的团队协作精神； 2）鼓励学生深入挖掘专业内容中的思政元素
课时：4个课时	课时：2个课时

在课堂的理论知识授课时，会穿插有关建筑行业的时事热点新闻以激发学生的兴趣。如在疫情防控期间，给学生讲解火神山、雷神山医院的建设与施工；在讲解建筑及设备图纸的同时，播放两个医院建设的录像，让学生切实感受到"中国速度"和"中国实力"，从而认识到新时代中国特色社会主义制度的优越性和蕴藏的党领导下人民群众应对危机时团结一致、奋不顾身的优良传统，从而坚定学生对中国特色社会主义的道路自信和制度自信。

"建筑体系整合"课程将原本的实践认知内容从2个课时增加到6个课时，这样就有充足时间到红色教育基地进行实地参观。被缩短的前2个课时是抽取建筑使用后整体评价的课堂知识，这部分内容被整合到实地参观中讲解，让学生能更好地理论联系实际。由于将原本的个人作业改成小组工作，这样不仅可以培养学生的团队合作精神，同时以组为单位可以缩短课堂汇报时间。

无论是"建筑体系整合"专业课的内容还是增添的课程思政内容，其教学内核都是教书育人，二者应是一个有机整体，从课程的多个方位和多个层面将专业知识传授和思想价值引领互相渗透起来，将立德树人作为教学主线，把专业课内容贯穿在这一主线里，通过激发学生的探究兴趣和学习热情，才能将"教、学、思"三者有机地结合起来。

4. 课程思政教育的教学实施

"建筑体系整合"采用了许多新型教学模式，如线上线下混合教学、翻转课堂等，如何将课程思政内容应用在新型教学模式，让这些优良的文化传统和思想品德真正地在学生身上内化于心，外化于行，这需要教师结合当代学生的特点，探索学生乐于接受的教学方法。

笔者利用新型教学模式，积极开展"走出去、领进来"的主动教学，尝试建构一个以学生的需要和兴趣为导向，以培养建筑师职业修养和正确价值观为目标的开放式、双向式的灵活教学模式。下面以"建筑体系整合"的实践认知教学内容为例，简述教学实施过程：

1）选择南京本地的红色文化资源，将学生组织到南京晨光1865创意产业园进行参观学习。教师带领学生们实地参观晨光1865园区不同时代的工业建筑，学习由于受不同时代建造技术和材料的影响，建筑物内不同的结构与构造、保温与隔热、采光等物理环境，探讨体系整合理论与建筑实践的关系。而这些历经沧桑的建筑，则用无声的语言告诉学生中国近现代工业建筑技术的发展，展示了中华民族军工和航天事业的崛起和发展。

2）引导学生选择金陵兵工展览馆，并以这个清代光绪年间建造的老厂房（2005年晨光集团进行了修缮）为研究对象，进行了各种室内空间、声、光、热、室内空气质量等一系列测量，对其改建成为展览馆后做建筑使用后整体评价。在进行建筑物理环境实地测量和评估时，金陵兵工展览馆中带有革命意义和历史传承的一幅幅图片、一件件实物给同学们留下了深刻的印象，让他们学习掌握建筑使用后整体评价理论知识的同时，也在潜移默化中接受了爱国主义教育。

3）聘请原晨光厂的退休老党员宋建中先生作为实践导师，从不同角度介绍了作为中国军事工业和兵器工业的摇篮"金陵兵工厂"，如何发展到今天的以国为重、以军为本的晨光集团，以及晨光集团为国防建设和航天事业所做出的重要贡献。

我国古人早就提倡"读万卷书、行万里路"。"课程体系整合"有意识地增加大量的综合实践内容，充分利用了南京本地的红色文化资源，带领学生走出教室，通过在有红色教育意义的实地学习建筑技术理论知识，在实地积极开展使用后建筑整体评估，让学生能够做到理论联系实际，充分了解我国建筑业的发展和现状。同时聘请老党员进行现场思政教学，让学生在学习建筑体系整合专业知识的同时，深刻了解当地的历史和文化，潜移默化地进行思政教育。

5. 课程思政教育的考核模式

学生思政水平的考核不同于专业知识的考核，更多的是体现在学生日常的精神风貌、言行举止上，如果简单地将思政课题指标化，在课业考试中加入思政试题，显然违背了"课程思政"的初衷和目标，效果也不会尽如人意。

为配合课程思政教改需要，学生的考核模式也做了相应调整，取消了传统的考试，取而代之的为平时表现和课程报告两大部分，特别是平时表现成绩提高到40%，包括学生出勤率、课堂讨论的参与程度、网络讨论的参与程度等，都会计算在内。课程报告则占总成绩的60%，采用实地建筑评估报告模式，让学生将理论知识应用在建筑实践中，并且建筑实地评估报告采用的是小组作业形式，以培养学生团队合作、顽强拼搏、尊重对手、尊重规则、不惧失败等价值品性。

在考核评分中，教学团队会明确告诉学生，无论是平时表现还是最终的课程报告，都会适当加入思政内容的权重评分，这样可以让学生有意识地学习课程思政内容，提高自己的政治理论水平。从学生递交的课程报告可以看出，添加权重评分项的措施极大地激励学生的思政学习，如最终的课程报告和课堂汇报，所有学习小组都专门添加了相关思政内容，除去专业知识，学生都或多或少地介绍了实地参观的历史背景，从"金陵兵工厂"的革命历史，到1949年以后，晨光集团又是如何以国为重、以军为本，为国防建设和航天事业做出重要贡献，这些汇报内容也让教师受到不少教育，真正做到了教学相长。

6. 课程思政建设的经验与思考

总体而言，"建筑体系整合"课程的思政改革实践较为成功，该课程已被南京大学研究生院评为研究生课程思政标案课程，并被推荐为南京大学研究生思政课程推荐案例。尽管在专业知识中增添了大量思政教学内容，在近三年的教改实践中，学生对该课程的评价反而提高了很多，很多学生认为所添加的教学知识可以帮助他们了解我国的国情和社情，对于他们将来的工作都非常有用。在对学生的回访中，大部分学生表示除去实地参观和实习之外，并没有感觉到明显的思政元素，但所有学生都表示实地参观学习的机会非常难得和有趣，不仅学到了很多的专业知识，也了解当地的社会文化和历史文脉，认识到一个建筑从设计、施工到使用，不仅面临着一系列综合、复杂的技术问题，也与社会环境息息相关。通过"建筑体系整合"较

为成功的思政改革实践，笔者认为有两点经验值得分享：

1）在重新建构课程体系基础上开展课程思政教育。"建筑体系整合"是一门专业性极强的研究生课程，其教学内容采用国外先进的专业技术理论和知识，如果简单粗暴地将课程思政内容添加在专业知识里，不仅会影响到课程本身的完整性和连贯性，还会影响到学生的学习兴趣，因此课程思政建设不但不会有正面效果，有时还会产生负面效应。笔者认为要想更好地开展课程思政建设，需要教师从多方面、多方位，甚至是多年的教学经验出发，在重新建构课程体系的基础上开展课程思政建设。特别是研究生在心理、生理年龄上都趋于成熟，大多数研究生已经形成了较稳定的价值观，他们不会简单地"认同"或"不认同"某个观点，而是把价值取向放在义利并重的考量中[9]，因此只有在重新建构课程体系中添加思政教育内容才能有效吸引学生的兴趣、有效契合研究生的价值诉求。如教学团队会针对建筑学研究生所面对的就业需求，着重介绍我国当前的建筑发展现状，创新创业的机会和挑战，以配合学生的实际需求。同时，走出课堂，开展实地参观则可以提高学生的学习兴趣，开阔眼界，更好地做到理论联系实际。

2）要将价值塑造的要素有机地融入专业知识。客观而言，不同课程开展"课程思政"的难易程度有很大差别，文科课程资源丰富，相对容易融合，理工科课程的设计难度相对加大[10-11]。习近平总书记早就指出："好的思想政治工作应该像盐，但不能光吃盐，最好的方式是将盐溶解到各种食物中自然而然吸收。"[12] 要想将课程思政内容有机融入专业课程教学，笔者认为需要将课程思政内容内化于专业课程的教学规划、课堂教授、学术研和评价体系中，千万不能将思想元素以思政课的教学模式直白地灌输给学生，而是在专业课学习中，用主流德育理念和科学的思维方法去引导学生，使其在获得知识的同时，引发价值领域的思考。并且有必要通过提高教师的自我修养，如有意识地加强自我的师德师风建设，通过自己的言传身教，让学生耳濡目染地学习到我们的理想信念、道德情操、扎实的学识和仁爱之心。

除去课程本身的思政建设，教学团队认为学校层面的支持也非常重要：首先，搭建一个适合课程思政教育的实践平台和整体课程体系。我国大学以前主要重视的是专业教育，而思政教学则是一个德育教育，这是一个潜移默化的长期教育过程。因此，在硬件上高校应该大力支持和建设可以覆盖整个专业教育的课程思政平台和交流平台，如与大学接轨的红色教育基地。其次，需要学校在政策和教学资源上的大力支持。课程思政建设对教师的主观能动性要求很高，不仅需要教师教学经验丰富，还需要教师大量时间和精力对课程内容和进程进行重新构建，添加不少新的课程内容。这需要学校在政策、教学资源和资金上的大力支持，如配备助教、对教师绩效考核的倾向。最后，思政课程建设必须建立健全组织领导体系，形成落实到院系领导的专项负责制度，因为每一门课程在课程思政建设上都会有不同需求。如"建筑体系整合"课程是春季开课，开课时间是1~8周，而南京大学研究生院的课程思

政建设项目则是从第8周才正式启动，而此时该课程已经基本结束。本次教改得益于南京大学建筑与城市规划学院党委书记的多次催办和特事特办，才真正开展起来。由此可见，课程思政的建设需要学校、领导、教师等多方联动，积极配合，才能真正开动起来，毕竟课程思政并不是简单地添加一些思政内容，而是以人为本地对课程教学结构进行根本变革。

　　中国建筑师的职业教育和职业修养，有必要加入"立德树人"的新元素。而践行"为党育人、为国育才"的历史使命，紧扣立德树人这一根本任务，把思想政治教育贯穿人才培养体系中，是每一位高校教师的应尽之则。依托研究生专业教育，促进专业教育与思政内容有机融合，让学生建立正确的人生观和价值观，可以有效弥补传统建筑学专业教育中的薄弱之处，才能培养出德才兼备的合格建筑师。尽管现阶段我国"课程思政"的教学改革主要在本科教育阶段推进，作为高等教育的更高层次，研究生思想政治教育也是高校思想政治教育工作的重要组成部分。由于研究生教学在培养目标和培养体系上都与本科教育有着很大区别，因此其课程思政建设也应与本科教育有所不同。"建筑体系整合"的课程思政教改是在重新建构课程体系基础上，重新定位课程，从课程内容、教学模式和考核方式等多个渠道、多个层面，潜移默化地引导学生建立正确的人生观和价值观；并从吸引学生学习兴趣出发，结合他们的实际需要，让研究生从复杂多变的客观实际中，亲身体会和明确自己的发展目标，建立运用中国理论分析中国问题的能力。教师则需要把握高等学校教书育人工作中价值、知识、能力这三个基本要素之间的关系，落实立德树人这一根本任务，这对促进学生更深层次、更多维度地理解建筑设计和城市规划具有积极促进作用。

In order to thoroughly implement General Secretary Xi Jinping's important exposition on education and the spirit of the National Education Conference, the *Several Opinions on Deepening the Reform and Innovation of Ideological and Political Theory Courses in Schools in the New Era* issued by the General Office of the CPC Central Committee and the General Office of the State Council, and the spirit of the *Guidelines for Ideological and Political Construction of Courses in Colleges and Universities* issued by the Ministry of Education, various universities in China have begun to comprehensively promote the ideological and political construction of courses in recent years[1–3]. Architecture is an interdisciplinary profession facing a series of comprehensive and complex technical and social issues. For the new generation of architects, it is not only necessary to strengthen vocational education, but also to cultivate correct values and professionalism, which must keep up with the pace of the times. As Qiu Kaijin pointed out, "Ideological and political construction of courses can help cultivate a correct outlook on life and values."[4] Only by establishing a correct outlook on life and values, can we cultivate qualified architects with both morality and talents.

At present, the reform of ideological and political education courses in China is mainly concentrated on the undergraduate educational stage, and it just started on graduate students. Although graduate study still falls under the category of vocational education, there are still significant differences compared with undergraduate education[5]. Graduate education emphasizes the cultivation of cognitive abilities, and the psychological and physiological age of graduate students are more mature than those of undergraduate students[6]. Therefore, simply applying the experience of ideological and political construction of undergraduate courses can lead to many problems. Previous studies have shown that there is a dilemma in the ideological and political education reform of graduate courses in terms of explicit, formalistic, and utilitarian teaching methods. Therefore, it is necessary to explore an effective development approach for the ideological and political construction of graduate courses based on the spiritual characteristics and value pursuit of the graduate group, combined with the training objectives and methods of graduate students. Taking the opportunity of the ideological and political teaching reform of graduate courses actively promoted by the Graduate School of Nanjing University, and taking the "Building System Integration" as an example, which is a graduate course of architectural technology taught by the author, how to integrate the ideological and political construction of the course into graduate teaching was studied in this paper from four aspects: the course orientation, teaching content, teaching mode, and assessment method. Based on sharing experience and lessons in the ideological and political education reform, more operational and practical graduate courses were explored for ideological and political education.

1. The difficulties faced by the ideological and political construction of the course of "Building System Integration"

"Building System Integration" is a postgraduate course offered by the School of Architecture and Urban Planning of Nanjing University. Its main content covers three parts of knowledge: building system integration theory, post-occupancy evaluation and total building performance, and research methods of building technology theory. The main teaching goal of this course is to enable students to discover and analyze the actual problems in the existing buildings by using total building performance methods, to guide students to re-examine and evaluate the concept and value system of architectural design, and establish building environmental awareness and sustainable development values. Although this course is compulsory for graduate students in the architectural technology direction of the School of Architecture and Urban Planning of Nanjing University, it is also open to all graduate students in Nanjing University for elective.

Although the overall teaching goal of the "Building System Integration" course is consistent with the spirit of Several Opinions on Deepening the Reform and Innovation of Ideological and Political Theory Courses in Schools in the New Era issued by the General Office of the CPC Central Committee and the General Office of the State Council, as a highly specialized engineering course, it faces the following three difficulties in implementing the ideological and political education reform of the course:

Firstly, "Building System Integration" is a new course on teaching content. It learns from National University of Singapore, the Chinese University of Hong Kong, Cornell University, Massachusetts Institute of Technology and other universities. These teaching contents cover a lot of foreign cutting-edge theoretical knowledge of architecture, contains a lot of advanced professional, theory and knowledge. At present, only the Department of Architecture of Nanjing University offers this course. Its main professional textbooks, handouts and reading materials are in English, and only a small part of the teaching contents have been translated into Chinese, so it is not easy to explore the ideological and political elements of the course.

Secondly, a relatively new teaching model has been adopted. This course is not only a required course for graduate students majoring in architectural technology, but also a graduate course for all graduate students of the School of Architecture and Urban Planning, even the whole university. Graduate students who choose the course come from all departments of Nanjing University. Some graduate students majoring in science and arts, such as Mathematics, Economics and Management, choose this course, beside of some traditional engineering departments. The students have different backgrounds and different level of knowledge. In order to cooperate with graduate students from different backgrounds, many new teaching models have been adopted, such as online and offline mixed teaching combined with MOOC, and flipped classroom, etc. However, there is currently limited research on how to apply ideological and political education of the course to new teaching models[7].

Finally, it comes to teaching schedules and teaching arrangements. The School of Architecture and Urban Planning of Nanjing University offers many graduate courses, so the teaching arrangement is very compact. For example, the course of "Building System Integration" is arranged in the first eight weeks of the semester, and each class lasts four hours. The abundance of course contents, compact course arrangements, and long class time each time pose a challenge for both students and teachers, especially for students taking it as the elective course from other departments.

Although the "Building System Integration" has borrowed from the training models of advanced Western countries in both the course content and teaching mode, regardless of the change, the four basic elements of its course positioning, teaching content, teaching mode, and assessment method dominate the overall construction of the course. Therefore, it is also necessary to start from these four aspects to carry out the ideological and political construction of the course of "Building System Integration".

2. Emphasize the ideological and political content in the course positioning

The so-called ideological and political construction of the course is an educational reform practice that starts from the overall ideological and political pattern. It is to integrate the educational goal of cultivating virtues and morality into all aspects of the course system, stimulate ideological and political elements in the course, combine knowledge impartation with value guidance, and guide students to develop comprehensively towards morality, intelligence, physical fitness, aesthetics, and labor[8]. The key and starting point of ideological and political education in the course is the course identity. It is rebuit the objectives and content of teaching, reconstruct the course system, and expand new educational connotations to integrate graduate professional education with ideological and political education. Only then can students learn professional knowledge while also grasping correct values and worldviews.

The education of architecture majors aims at cultivating future architects, but there are many problems in the field of architecture education in China today. For example, the professional teaching model led by architectural design often results in architectural students placing more emphasis on art and less emphasis on technology, emphasizing design form analysis and less emphasis on objective quantification, and a disconnect between theory and practice. This leads to the lack of firm thinking and understanding of China national and social conditions among young architects. Therefore, there are many chaos, such as "greed for grandeur, workshop the western culture, and strangeness". On October 15, 2014, when General Secretary of the Central Committee of CPC, Chairman of China, Chairman of the Central Military Commission Xi Jinping hosted a symposium on literary and artistic work in Beijing, he stated that "strange and bizarre buildings" should not be encouraged.

For these issues, the positioning of the ideological and political construction of the course of "Building System Integration" starts from two breakthroughs. Firstly, it breaks the barriers between the professional theory of architectural technology and the cultivation of architects' outlook on values and life. Secondly, it breaks the barriers between the transmission of professional knowledge and the cultivation of ideological education. It attempts to integrate the spirit of moral education into professional course education, guide students to turn theory into practice and practice knowledge into virtues, become a true socialist builder. The author repositioned the course of "Building System Integration" by breaking these two barriers, on the premise of learning from the advanced western knowledge, to implement China new era architectural policy of "applicability, economy, green and beauty". Meanwhile, it strengthens cultural self-confidence, enhances cultural self-awareness, continues the Chinese cultural context, it not only shows the zeitgeist, and also learn to reflect regional cultural characteristics. Meanwhile, the course should also strengthen the ideological education of students from various aspects, especially stimulating the scientific spirit of architectural students, and cultivating their professionalism.

3. Reconstructing and integrating the teaching content of "Building System Integration"

The contents of "Building System Integration" includes three parts: the basic theory of building technology, the theory of building system integration, and the post-occupancy evaluation and total building performance. The basic theory of building technology is taught by the author in MOOC, which is not easily modified. The offline teaching first explains the theory of building system integration, which includes what building system integration is, and how to interpret various architectural systems from the perspective of integration, as well as the mutual influence and constraint relationships between these systems. The course next introduces the theory and methods of post-occupancy evaluation and total building performance, and how to use the concept of building integration to analyze and evaluate the overall environment of buildings.

According to the new course identity, the author has restructured and integrated the main contents of offline teaching. The comparison of the new and old course contents and the class hours is shown in Table 1. The course content revolves around value guidance and moral cultivation for students. For example, at the beginning of the course, we used to only focus on the development of western architecture after the Industrial Revolution when teaching the history of architectural technology development. However, now we incorporate the excellent ancient architecture and advanced technology that have emerged in the 5 000 years history of China, especially how Chinese residential buildings follow the harmonious relationship between architecture and environment, architecture and people, architecture and society in many aspects, such as site selection, spatial layouts, and construction technology.

In order to make the course of "Building System Integration" truly localized, when teaching the basic theory of building system integration, we consciously introduce it in combination with the latest national and social conditions of our country, from the latest *Opinions on Promoting Green Development in Urban and Rural Construction* by the General Office of the CPC Central Committee and the General Office of the State Council to the development strategy of the 14th Five-Year Plan for China, so that students can learn the concept of ecological priority and green development. In addition, while allowing students to learn the theory of post-occupancy evaluation, it is also proposed to re-examine the current development status of architecture in China from an ecological and rational perspective, comparing the ingenuity in handling the relationship between humans and nature, humans and humans, and humans and resources in traditional Chinese houses, and explore the simple ecological architectural ideas and humanistic awareness and harmonious concepts embedded in them, to promote, improve, and develop modern architectural culture, support and form healthy and scientific modern architectural cultural concepts and practices.

Table 1 Comparison of teaching content and hours before and after the ideological and political education reform of the course

Original teaching content	The teaching content after the ideological and political reform of the course
Beauty and Art in Architectural Systems: A History of Western Architectural Technology Development after the Industrial Revolution	Increase the outstanding ancient buildings and the glorious achievements of ancient people in construction technology that have emerged in the 5,000 years' history of China, as well as the thriving development status of Chinses building intelligence level today
Lesson: 4 class hours	Lesson: 4 class hours
Theory of building system integration	Based on the original theoretical knowledge, add various current building codes and the development trend of Chinese buildings, from the latest *Opinions on Promoting Green Development in Urban and Rural Construction* by the General Office of the CPC Central Committee and the General Office of the State Council to the development strategy of the 14th Five-Year Plan for China, let students learn the concept of ecological priority and green development from these latest national and social conditions

Continued

Original teaching content	The teaching content after the ideological and political reform of the course
Lesson: 10 class hours	Lesson: 10 class hours
Theory of post-occupancy evaluation and total building performance	Let students learn to understand architecture from the perspective of building users and with a people-oriented approach. Let students start more from the actual use of the building, consider planning and architectural design from the overall architectural environment and sustainable development, rather than just from spatial use and artistic design
Lesson: 12 class hours	Lesson: 12 class hours
Practical cognition	The practical cognition has been increased from the original 2 class hours to 6 class hours, making full use of the local red cultural resource in Nanjing, bringing students to the red education base, where a series of measurements of indoor space, sound, light, heat, indoor air quality, etc. have been carried out, and the overall assessment of the buildings after use has been carried out
Lesson: 2 class hours	Lesson: 6 class hours
Student report	1) Transforming individual assignments into group assignments to cultivate students' teamwork spirit; 2) Encourage students to deeply explore the ideological and political elements in their professional content
Lesson: 4 class hours	Lesson: 2 class hours

News about current events in the construction industry will be interspersed to attract students' interest during the teaching of theoretical knowledge in the classroom. For example, students were told about the construction of Huo-shen-shan and Lei-shen-shan Hospital during the epidemic. Meanwhile, the video of the construction of the two hospitals was shown to the students while explaining the drawings of buildings and equipments, so that students could really feel the "speed of China" and "strength of China", so as to recognize the advantages of the Socialism with Chinese characteristics system in the new era, and the fine tradition of solidarity and recklessness of the people under the leadership of the Party in the face of crisis, to strengthen students' confidence in the path and system of Socialism with Chinese characteristics.

The course of "Building System Integration" has increased the original practical cognitive content from two to six class hours, providing ample time for on-site visits to the red education site. The first 2 shortened class hours were based on classroom knowledge extracted from the post-occupancy evaluation, which was integrated into on-site visits, enabling students to better integrate theory with practice. By changing the original individual assignment to group work, not only can the teamwork spirit of students be cultivated, but also group based reporting time can be shortened.

Whether it is the professional content of the course of "Building System Integration" or the added ideological and political content of the course, the teaching core is to teach and educate people. The two should be an organic whole, permeating the teaching of professional knowledge and the guidance of ideological values from multiple aspects and levels of the course, taking moral education and talent cultivation as the main teaching line, and integrating the professional course content into this main line, by stimulating students' exploration interest and learning enthusiasm, Only then can we organically combine "teaching, learning, and thinking".

4. Teaching implementation of ideological and political education in course

"Building System Integration" has adopted many new teaching modes, such as online and offline mixed teaching, Flipped classroom, etc. How to apply the ideological and political content of the course to the new teaching modes, so that these excellent cultural traditions and ideological and moral qualities can be truly internalized in students and externalized in practice, which requires teachers to combine the characteristics of contemporary students and explore teaching methods that students are willing to accept.

The author utilized the new teaching modes to carry out the active teaching mode of "going out and bringing in", and attempts to construct an open and two-way flexible teaching mode based on students' needs and interests, with the goal of cultivating architects' ethics and correct values. Taking the practical cognition teaching content of "Building System Integration" as an example, the following briefly describes the teaching implementation process:

1) The local red cultural site in Nanjing was selected and students were organized to visit and study in Nanjing M&G 1865 Creative Industrial Park. The teacher led the students to visit the industrial buildings built in different eras, learning different building technologies such as structure and construction, insulation and heat preservation, and lighting effects due to the influence of materials and construction technology at different times. They also explored the relationship between theory of system integration and practice of building. These buildings, which have gone through vicissitudes of life, tell students the development of modern industrial building technology in China in silent language, and showed the rise and development of China's national military industry and aerospace industry.

2) Students were guided to choose the Jinling Military Industry Exhibition Hall. This old factory building built during the Guangxu period of the Qing Dynasty (repaired by M&G Group in 2005) is selected as the research object. A series of measurements, such as sound, lighting, heating, indoor air quality, etc. were conducted. The students conducted an overall evaluation of the building after transforming it into an exhibition hall. During the on-site measurement and evaluation of the physical environment of buildings, the pictures and objects with revolutionary significance and historical heritage in the Jinling Military Industry Exhibition Hall left a deep impression on the students, allowing them to learn and master the theoretical knowledge of post-occupancy evaluation and total building performance, while they also subtly received patriotic education.

3) Song Jianzhong, a retired veteran Party member of the former M&G Factory, was employed as a practical tutor to introduce from different perspectives how the "Jinling Arsenal", the cradle of China's military industry and weapons industry, has developed into today's M&G Group, which focuses on the country and the military, and how M&G Group has made important contributions to national defense construction and aerospace industry.

The ancients of our country used to advocate "reading thousands of books and traveling thousands of miles". "Building System Integration" consciously increased a large number of comprehensive practical content, made full use of the local red

Cultural resources in Nanjing, took students out of the classroom, and carried out POE study on the red education base. As a result, students could integrate theory with practice, as well as fully understand the development and current situation of China's construction industry. At the same time, veteran Party members were hired to conduct on-site ideological and political education, allowing students to have a deep understanding of local history and culture while learning professional knowledge about building system integration, and to subtly carry out ideological and political education.

5. The assessment model of ideological and political education of the course

The assessment of students' ideological and political level is different from the assessment of professional knowledge, which is more reflected in their spirit and appearance, words and actions. If we simply index the ideological and political topics and include ideological and political test questions in the academic exam, it clearly goes against the original intention and goal of, and the effect will not be satisfactory.

In order to meet the needs of ideological and political education reform of the course, the assessment mode of students has also been adjusted accordingly, eliminating the traditional exam and replacing it with two parts: daily performance and the course report. Especially, the daily performance score has been increased to 40%, including attendance rate, participation in classroom discussion, and participation in online discussion, etc.. The course report accounts for 60% of the total score, which use the on-site architectural evaluation report to allow students to apply theoretical knowledge to architectural practice. The on-site building evaluation report also adopts the form of group assignments to cultivate students' values and qualities such as teamwork, tenacious struggle, respect for opponents, respect for rules, and fearlessness of failure.

Before given the assessment scoring, the teaching team would clearly inform students that both their daily performance and the final course report will be appropriately weighted with ideological and political content.This method can help students consciously learn the ideological and political content and improve their political theory level. From the course reports submitted by students, the measures of adding weighted scoring items have greatly stimulated students' ideological and political learning. For example, in the final reports, all students groups added relevant ideological and political content in their final report specifically. In addition to professional knowledge, students introduced the historical background of the field visits. From the revolutionary history of the "Jinling Arsenal" to the founding of PRC, how does the M&G Group focus on the country taking the military as an foundation and making important contributions to national defense construction and the aerospace industry. These reports have also gave teachers some new idea, which truly achieving mutual benefit between teaching and learning.

6. Experience and reflection on ideological and political construction of the course

In general, the practice of ideological and political reform of the course "Building System Integration" is relatively successful. This course has been rated as an outstanding course of ideological and political education for graduate students by the Graduate School of Nanjing University, and suggested as a recommended case of ideological and political education for graduate students of Nanjing University. Although a large amount of ideological and political teaching content has been added to professional knowledge, during the past three years of teaching reform practice, students have improved their evaluation of the course. Many students now believe that the added teaching knowledge can help them understand the national and social conditions of our country, which is very useful for their future work. In the follow-up students, most of them stated that apart from on-site visits and internships, they did not feel any obvious ideological and political elements. However, all students expressed that the opportunities for on-site visits and learning were very rare and interesting. They not only learned a lot of professional knowledge, but also understood the local social culture and historical contexts. They realized that a building, from design, construction to use, not only faces a series of comprehensive and complex technical problems, It is also closely related to the social environment. Through the successful practice of ideological and political reform through the "Building System Integration", the author believes that there are two points worth sharing:

1) Based on reconstructing the course, ideological and political education will be carried out in the course. "Building System Integration" is a highly specialized graduate course that adopts advanced professional technical theories and knowledge from abroad. If the ideological and political content of the course is simply and roughly added to the professional knowledge, it will not only affect the integrity and coherence of the course itself, but also affect students'

learning interests. Therefore, the ideological and political construction will not have positive effects, but sometimes have negative effects. The author believes that teachers need to start from multiple aspects, multiple directions, and even years of teaching experience, and carry out ideological and political construction based on reconstructing the course system to better promote ideological and political construction. Especially graduate students tend to mature in terms of psychological and physiological age, and most of them have formed relatively stable values. They do not simply "agree" or "disagree" with a certain viewpoint, but instead place value orientation in the consideration of both righteousness and benefit[9]. Therefore, only by adding the ideological and political education content in the reconstruction of the course system can effectively attracted student' interests and met the value demands of graduate students. For example, in response to the employment needs faced by graduate students in architecture, the teaching team would focus on introducing the current development status of architecture in China, opportunities and challenges for innovation and entrepreneurship. The key is, to meet the actual needs of students. Meanwhile, going out of the classroom and conducting on-site visits can enhance students' interest in learning, broaden their outlook, and better integrate theory with practice.

2) To organically integrate the elements of value shaping into professional knowledge. Objectively, the difficulties of carrying out "ideological and political courses" varies greatly among different courses[10-11]. The resources of liberal arts courses are abundant and relatively easy to integrate, while the design difficulty of science and engineering courses is relatively increased. General Secretary Xi Jinping pointed out early on: "Good ideological and political work should be like salt, but not just eating salt. The best way is to dissolve salt into various foods and naturally absorb it"[12]. We must not impart ideological elements directly to students through the teaching mode of ideological and political courses. Instead, we should use mainstream moral education concepts and scientific thinking methods to guide students in professional course learning, while acquiring knowledge and triggering thinking in the field of value. And it is necessary to improve the self-cultivation of teachers, such as consciously strengthening the construction of their own professional ethics and conduct. Through their own words and deeds, students can learn from our ideals, beliefs, moral sentiments, solid knowledge, and benevolence.

Apart from the course itself, the teaching team believes that support from school level is also very important: Firstly, it is necessary to build a practical platform and overall course system suitable for ideological and political education. In the past, Chinese universities mainly focused on professional education, while ideological and political education was moral education, which was a imperceptible and long-term educational process. Therefore, in terms of hardware, universities should vigorously support and build ideological and political course platforms and communication platforms that can cover the entire professional education, such as red education sites that are in line with universities. Secondly, it requires strong support from schools in terms of policies and teaching resources. The ideological and political construction of the course requires high subjective initiative from teachers. Teachers not only need rich teaching experience, but also a lot of time and energy to reconstruct the course content and process, adding many new course materials. These requires strong support from schools in terms of policies, teaching resources, and funding, such as the provision of teaching assistants and a tendency towards teacher performance evaluation. Finally, the construction of ideological and political courses must establish a sound organizational leadership system, becoming a special responsibility system implemented by department leaders, because each course has different needs in the construction of ideological and political courses. For example, the course of "Building System Integration" starts in the first week of spring semester and lasts for eight weeks, while the graduate school of Nanjing University was officially launched the ideological and political construction project from the 8th week. the course has basically ended at that time. It is need to thanks the Secretary of the Party Committee of the School of Architecture and Urban Planning of Nanjing University. His repeated urging and special handling make this project carry out successfully. From this, it can be seen that the construction of ideological and political education of the course requires the active cooperation and collaboration of schools, officials, teachers, and other parties in order to truly start. After all, course ideological and political education is not simply about adding some ideological and political contents, but rather a fundamental transformation of the course teaching structure based on people.

It is necessary to incorporate new elements of "cultivating virtues and morality" into the vocational education and cultivation of Chinese architects. Practicing the historical mission of "educating people for the Party and China", closely adhering to the fundamental task of cultivating virtues and morality, and integrating ideological and political education into the talent cultivation system is the duty of each university teacher. Relying on graduate professional education to promote the organic integration of professional education and the ideological and political content, enabling students to establish correct outlooks on life and values, can effectively compensate for the weaknesses in traditional architecture professional education, and cultivate qualified architects with both morality and talents. Although the teaching reform of "ideological and political education of the course" in China at this stage is mainly promoted in the undergraduate educational stage, as a higher level of higher education, graduate education is also an important part of ideological and political education in colleges and universities. Due to the significant differences in training objectives and systems between graduate education and undergraduate education, the ideological and political construction should also be different. The ideological and political education reform of the "Building System Integration" is based on the reconstruction of the course system, repositioning the course, and imperceptible guiding students to establish a correct outlook on life and values from multiple channels and levels such as the course content, teaching mode, and assessment methods,etc.. Starting from attracting students' interest in learning, combined with their actual needs, graduate students can personally experience and clarify their development goals from complex and ever-changing objective reality, and establish the ability to apply Chinese theories to analyze Chinese problems. Teachers need to grasp the relationship between the three basic elements of value, knowledge, and ability in the teaching and education work of higher education institutions, and implement the fundamental task of cultivating virtues and morality. This has a positive effect on promoting students' deeper and more multidimensional understanding of architectural design and urban planning.

参考文献

[1] 王宇辉.建筑材料课程思政教育实践与探索[J].山西建筑,2021,47(16):191-192.
[2] 卢黎,谢强,朱正伟,等.工科专业课程思政教学方案设计探索与实践：以土力学课程为例[J].高等建筑教育,2021,30(3):108-113.
[3] 武震林,潘路军,韩秀友.本科基础课"光学"课程思政教学改革探索[J].教育教学论坛,2020,496(50):34-35.
[4] 邱开金.从思政课程到课程思政,路该怎样走[N].中国教育报,2017-03-21(10).
[5] 吴蔚.研究生建筑技术教学之"翻转课堂"[J].高等建筑教育,2017,26(2):36-39.
[6] 王茜."课程思政"融入研究生课程体系初探[J].研究生教育研究,2019(4):64-68,75.
[7] 陆道坤.新时代课程思政的研究进展、难点焦点及未来走向[J].新疆师范大学学报(哲学社会科学版),2022,43(3):43-58.
[8] 高德毅,宗爱东.从思政课程到课程思政：从战略高度构建高校思想政治教育课程体系[J].中国高等教育,2017(1):43-46.
[9] 顾晓英.创新思政课程 培育合格人才[J].思想政治工作研究,2017(1):23-24.
[10] 杨涵.从"思政课程"到"课程思政"：论上海高校思想政治理论课改革的切入点[J].扬州大学学报(高教研究版),2018,22(2):98-104.
[11] 刘笑吟,徐俊增.工科专业课程融合思政教育的探索与实践[J].高教学刊,2020(20):186-188,192.
[12] 李永胜.好的思想政治工作应该像盐[J].政工学刊,2017(11):76-77.

课程概览
COURSE OVERVIEW

建成环境导论与学科前沿
INTRODUCTION TO ARCHITECTURAL ENVIRONMENT AND FRONTIERS OF DISCIPLINES

丁沃沃
DING Wowo

课程介绍

建成环境是人类生产与生活的基本场所，是生存和发展的重要环境。建成环境的优劣关系到每个人的生存状况，建成环境的构建涉及多个学科，与其相关的各类知识是多个学科的共同基础。有史以来，一方面人们依靠技术进步不断从地球获取资源的同时也不断创造出更加高效和舒适的生存空间；另一方面，随着人们对自然界认知的更新，人们也在不断调整建构建成环境的方法和路径。

因此，从专业的角度了解建成环境的概况、进展、问题和前景对于刚刚踏入学科的初学者来说是后续学习的基础。此外，本课程力承担训练大学生学习方法的任务，以教学过程为载体，引导学生如何借助新的媒体技术获取知识，培养其独立思考、思辨的能力，促进学生尽快完成从中学到大学的学习方法的转型，为今后的学习打好基础。

课程要求

1. 理解随着社会转型，城市建筑的基本概念在建筑学核心理论中的地位以及认知的视角。

2. 通过理论的研读和案例分析理解建筑形式语言的成因和逻辑，并厘清中、西不同的发展脉络。

3. 通过研究案例的解析理解建筑形式语言的操作并掌握设计研究的方法。

Course descriptions

The built environment is the basic place for human production and life, and an important environment for survival and development. The quality of the built environment is related to everyone's living conditions. The construction of the built environment involves many disciplines, and various kinds of knowledge related to it are the common foundation of many disciplines. Historically, on the one hand, people rely on technological progress to constantly obtain resources from the earth, and also constantly create more efficient and comfortable living spaces; on the other hand, with the renewal of people's understanding of nature, they are constantly adjusting the methods and paths of constructing the built environment.

Therefore, understanding the general situation, progress, problems and prospects of the built environment from a professional perspective is basic for beginners who have just entered the discipline. In addition, this course aims to undertake the task of training university students on learning methods, using the teaching progress as a carrier to guide students on how to use new media technologies to acquire knowledge, cultivate their independent thinking and critical thinking abilities, and promote students to complete the transition of learning methods from high school to university as soon as possible, laying a solid foundation for future learning.

Course requirements

1. To understand the status and cognitive perspective of basic concepts of urban buildings in the core theory of architecture with the social transformation.

2. To understand the reason and logic of the architectural form language and different development processes in China and the West through theory reading and case analysis.

3. To understand the operation of the architectural form language and grasp methods of design and research by analyzing cases.

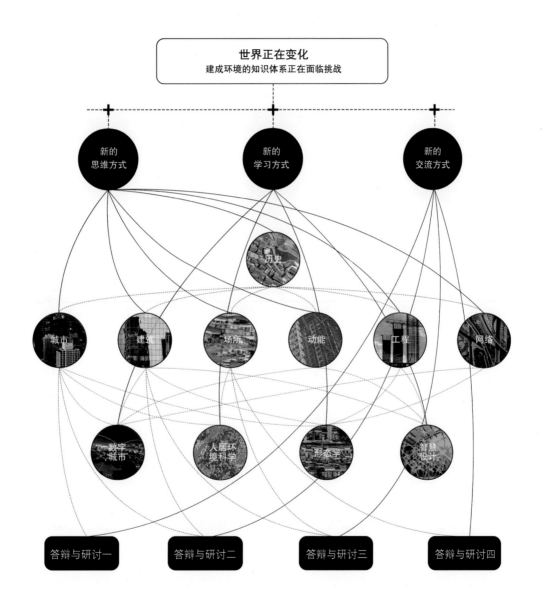

设计基础
DESIGN FOUNDATION

鲁安东　梁宇舒　刘铨　史文娟　赵潇欣　黄春晓
LU Andong, LIANG Yushu, LIU Quan, SHI Wenjuan, ZHAO Xiaoxin, HUANG Chunxiao

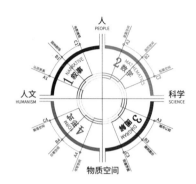

第一阶段：知觉、再现与设计
　　知识点：人与物——材料的知觉特征（不同状态下的色彩、纹理、平整度、透光性等）与物理化学特征（成分、质量、力学性能等）；摄影与图片编辑——构图与主题、光影与色彩；三视图与立面图绘制；排版及其工具——标题、字体、内容主次、参考线。

第二阶段：需求与设计
　　知识点：人与空间——空间与尺度的概念，行为、动作与一个基本空间单元或空间构件尺寸的关系；平面图、剖面图、轴测图的绘制；线型、线宽、图幅、图纸比例、比例尺、指北针、剖断符号、图名等的规范绘制。

第三阶段：制作与设计
　　知识点：物与空间——建构的概念；空间的支撑、包裹与施工；实体模型制作——简化的建造；计算机建模工具——虚拟建造；透视图绘制。

第四阶段：环境与设计
　　知识点：人、物与空间——城市形态要素、城市肌理与城市外部空间的概念；街道系统与交通流线；土地划分与功能分类；总平面图、环境分析图（图底关系、交通流线、功能分区、绿地景观系统）、照片融入表达。

Phase one: Perception, representation and design
Knowledge points: Humans and objects—the perceptual characteristics of materials (color, texture, flatness, light transmittance, etc. in different states) and physical and chemical characteristics (composition, quality, mechanical properties, etc.); photography and picture editing—composition and themes, light, shadow and color; three-view and elevation drawings; layout and its tools—headings, fonts, content priority, and reference lines.

Phase two: Demands and design
Knowledge points: Humans and space—the concept of the space and scale, the relationship between behavior, action and the size of a basic spatial unit or spatial component; plan, section, and axonometric drawings; the standard of the line type, width, map size, drawing scale, scale bar, compass, section symbol, drawing title etc.

Phase three: Production and design
Knowledge points: Objects and space—the concept of construction; space support, wrapping and construction; model making—simplified construction; computer modeling tools—virtual construction; perspective drawings practice.

Phase four: Environment and design
Knowledge points: Humans, objects and space—the concept of urban form elements, urban texture and urban external space; street system and traffic flow; land division and functional classification; general plan, environmental analysis map (relationship between the map and ground, traffic flow, functional zoning, green space landscape system), integration of images in expression.

人与物	人与空间	物与空间	人、物与空间
3周（个人作业） 每组10人左右	3周（个人作业） 每组10人左右	4周（个人作业） 每组10人左右	4周（个人作业） 每组10人左右
A1 鲁安东 理解身体	B1 鲁安东 场所认知	C1 鲁安东 园林剧场	D1 黄春晓 人与自然
		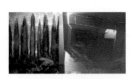	
A2 梁宇舒 回应动作的物	B2 梁宇舒 儿童空间单元设计	C2 梁宇舒 乡土住屋与结构演绎	D2 梁宇舒 乡土住屋与聚落重构

A3 刘铨 砖与砌块	B3 刘铨 楼梯间	C3 刘铨 观景塔设计	D3 刘铨 观景塔选址与总平面设计
	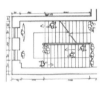		
A4 史文娟 行为再现与装置设计	B4 史文娟 独居公寓	C4 史文娟 快递站的设计与结构	D4 史文娟 校园小景观设计

A5 赵潇欣 家具与尺度	B5 赵潇欣 双层空间设计	C5 赵潇欣 建筑节点	D5 赵潇欣 大行宫地铁站分析与改造
			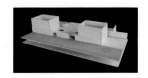

设计基础 DESIGN FOUDNATION

模块A: 回应动作的物
MODULE A: OBJECTS THAT RESPOND ACTION

梁宇舒
LIANG Yushu

课程目标
　　了解身体尺度；理解人与物的关系（人因工程学）；熟练使用度量与图示方法。

分组模块: 回应动作的"物"的设计

教学进程
第一周
　　摄像记录一套完整的动作过程，并用软件处理导出6张连续画面，根据人体结构学原理标注出人体结构线及身体作用力部位；进行人体运动画面叠加，并绘制分析图纸。
第二周
　　设计一个"物"（工具或家具），能够与人的身体发生的动作（第一周记录的动作）产生联系；手绘图纸：三视图及轴测图。
第三周
　　考虑了人的使用的物的设计表达。最终成果：独立完成A1图纸1张。

Course objectives
Learn body scales; grab understanding of relationships between humans and things (ergonomics); be expert in the use of measurement and diagramming methods.

Grouping module: Design of "objects" that respond to action

Teaching progress
Week 1
Take videos of a complete set of movement process, use software to export 6 consecutive pictures, and marks the human body structure line and force parts according to the principles of the body structure; superimpose human body movement pictures, and draw analysis drawings.
Week 2
Design an "object" (tools or furniture) that can be connected with the action of the human body (the action recorded in the first week); hand-drawn drawings: three-view and axonometric drawings.
Week 3
The design expression of objects that consider human use. Final result: Complete an A1 drawing independently.

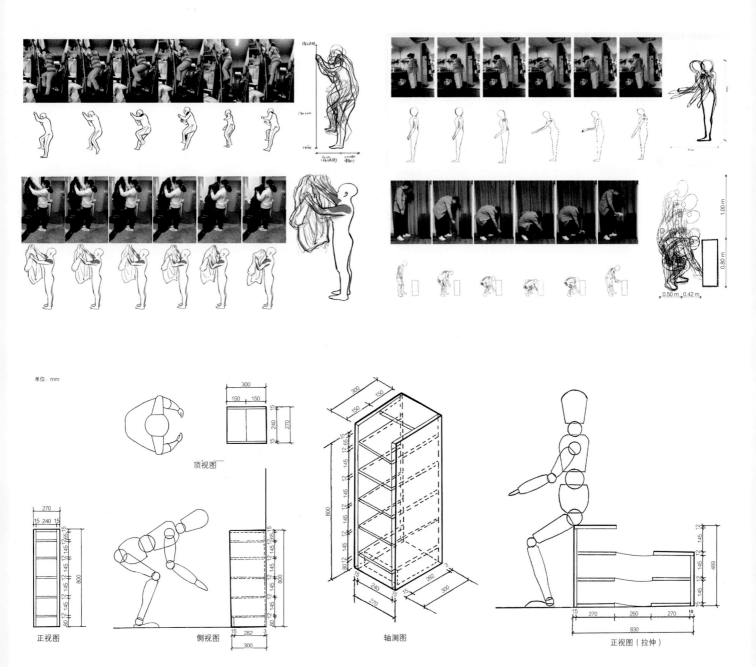

学生：赵同　Student: ZHAO Tong

设计基础 DESIGN FOUDNATION

模块B: 场所认知
MODULE B: PLACE COGNITION

鲁安东
LU Andong

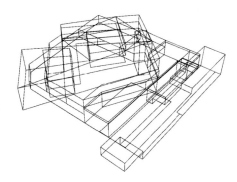

课程目标
　　理解功能；理解行为学（同时认识个体差异）；日常空间的使用场景；观察、记录与相应分析方法。

分组模块：场所认知

教学进程
第一周
　　自选一个校园内的日常空间（图书馆、自习室、教室、学生宿舍、咖啡厅、健身房等）进行测量、观察和记录（行为记录、环境/感知记录、事件记录），采用摄影、PPT、图表等形式进行汇报。
第二周
　　优化制图；增加测绘、拼贴图等表达形式；绘制轴测图。
第三周
　　方案模型 + 图纸。
第四周
　　独立完成 A1 图纸 1 张，全面地认知、分析和表现 1 个校园场所。

Course objectives
Understanding functions; understanding behavior (while recognizing individual differences); usage scenarios of daily space; observation, recording and corresponding analysis methods;

Grouping module: Place awareness

Teaching progress
Week 1
Choose a daily space on campus (the library, study room, classroom, student dormitory, coffee shop, gym, etc.) to measure, observe and record (behavior records, environment/perception records, event records), using photography, PPT, charts etc. to report.
Week 2
Optimize drawings; add mapping, collage and other forms of expression; draw axonometric drawings.
Week 3
Scheme model + drawings.
Week 4
Independently complete one A1 drawing to comprehensively recognize, analyze and represent a campus location.

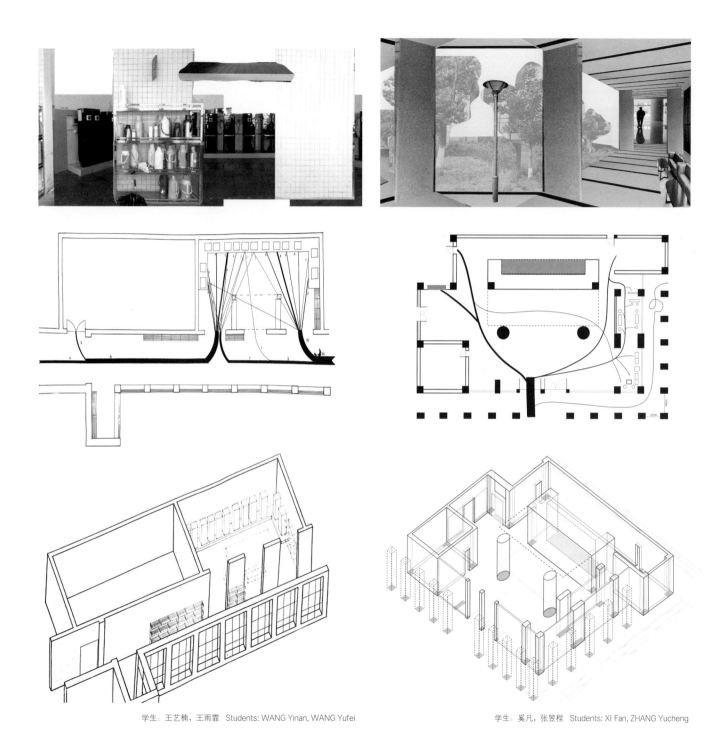

学生：王艺楠，王雨霏　Students: WANG Yinan, WANG Yufei

学生：奚凡，张昱程　Students: XI Fan, ZHANG Yucheng

设计基础 DESIGN FOUDNATION

模块C: 园林剧场
MODULE C: GARDEN THEATER

鲁安东
LU Andong

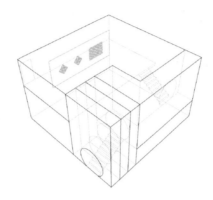

课程目标
理解真实建筑物（实体建造与虚体空间的互相作用）；材料、结构；图纸、模型与真实空间的对应关系。

分组模块：园林剧场

教学进程
第一周
1. 自选四个园林场景片段（来自网络）组成一段带有起承转合的体验，并配以文字叙述。
2. 运用空间要素将四个场景转化为更有诗意的新体验。
3. 完成概念模型（空间关系）。

第二周
1. 设计一个空间容器，将四个空间场景搭建成一个立体空间。
2. 完成空间模型（1∶30）。
3. 完成平面图、剖面图的绘制（1∶50）。

第三周
1. 将空间模型深化为建筑模型（含建造体系，1∶30）。
2. 深化平面图、剖面图的绘制（1∶50）。

第四周
1. 独立完成A1图纸1张。
2. 整理平、立、剖面与轴测图以及模型照片，进行排版。

Course objectives
Catch on to real buildings (the interaction between physical construction and virtual space); materials, structures; the corresponding relationship between drawings, models and real space.

Grouping module: Garden theatre

Teaching progress
Week 1
1. Select four fragments of garden scenes (from the Internet) to form an experience with transitions, accompanied by text narrations.
2. Using spatial elements to transform the four scenes into a more poetic new experience.
3. Complete the conceptual model (of spatial relationship).

Week 2
1. Design a space container and integrate four space scenes into a three-dimensional space.
2. Complete the space model (1:30).
3. Complete the plan and section drawings (1:50).

Week 3
1. Deepen the spatial model into an architectural model (including the construction system, 1:30).
2. Refine the plan and section drawings (1:50).

Week 4
1. Independently complete an A1 drawing.
2. Organize plan, elevation, section and axonometric drawings and model photos for typesetting.

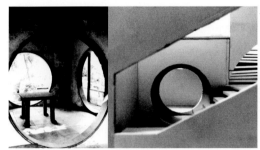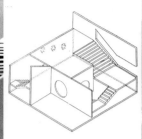

学生：周锦润　Student: ZHOU Jinrun

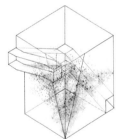

学生：郑锜泺　Student: ZHENG Qiluo

学生：王争鸣　Student: WANG Zhengming

设计基础 DESIGN FOUDNATION

模块D: 乡土住屋与聚落重构
MODULE D: RURAL HOUSING AND SETTLEMENT RECONSTRUCTION

梁宇舒
LIANG Yushu

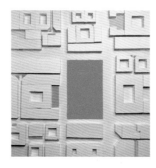

课程目标

理解更大尺度环境中的建筑（城市中或环境中的建筑）；认知建筑体与外部空间的互相影响；环境感知及其记录与分析方法。

分组模块：乡土住屋与聚落重构

教学进程
第一周
1. 从教师提供的人类住屋及聚落样本范围内，选取一个特定类型的典型聚落/村落，进行资料查找和形态分析；2. 绘制（整体/局部）聚落形态总平面分析图；3. 整理一篇PPT，内容包含所研究聚落对象的（资料）地理位置、自然环境、地形地貌；自绘聚落肌理分析图、道路分析图、景观要素分析图。
第二周
1. 在聚落组织逻辑分析基础上，确定一个聚落逻辑要素（如格子式、等高线式、中心式、轴线式、放射式、组团式、垂直式等）进行发展；2. 制定聚落中建筑单元体的标准模块及其变体，进行组合。计划块数，以及至少有一个社区中心（建议借助SketchUP软件）。3. 制作能够表达该聚落空间场所的简版建筑模型（表达2~3个模块之间的场所关系）。
第三周
1. 制作场地模型（40 cm×40 cm / 35 cm×50 cm），根据地理特征可以保留一定的景观要素。2. 基于特定的聚落结构要素和住屋单体原型，设计一座满足未来生活想象的"新型社区"。
第四周
1. 深化制作模型；2. 汇总最终成果并排版。

Course objectives

Understand buildings in a larger scale environment (buildings in cities or environments); recognize the interaction between buildings and external space; learn environmental perception and its recording and analysis methods.

Grouping module: Rural housing and settlement reconstruction

Teaching progress
Week 1
1. Select a specific type of typical settlements/villages from the sample range of human housing and settlements provided by the teacher, and conduct data search and morphological analysis; 2. Draw a (whole/partial) general plan analysis map of settlement morphologies; 3. Organize PPT, including (data) geographical location, natural environment, topography of the settlement object under study; Draw settlement texture analysis map, road analysis map, landscape element analysis map.
Week 2
1. Based on the logical analysis of the settlement organization, determine a logical element of the settlement (such as grid, contour, center, axis, radial, group, vertical, etc.) for development; 2. Formulate the settlement building module and combine its variants of the unit body. Plan the number of blocks, and with at least one community center (SketchUP software is recommended). 3. Make a simplified architectural model that expresses the spatial place of the settlement (expresses the relationship between 2–3 modules).
Week 3
1. Make a site model (40 cm × 40 cm/35 cm × 50 cm), and retain certain landscape elements according to geographical features. 2. Based on specific settlement structural elements and single housing prototypes, design a "new community" that meets the imagination of future life.
Week 4
1. Perfect the model; 2. Summarize the final results and typeset.

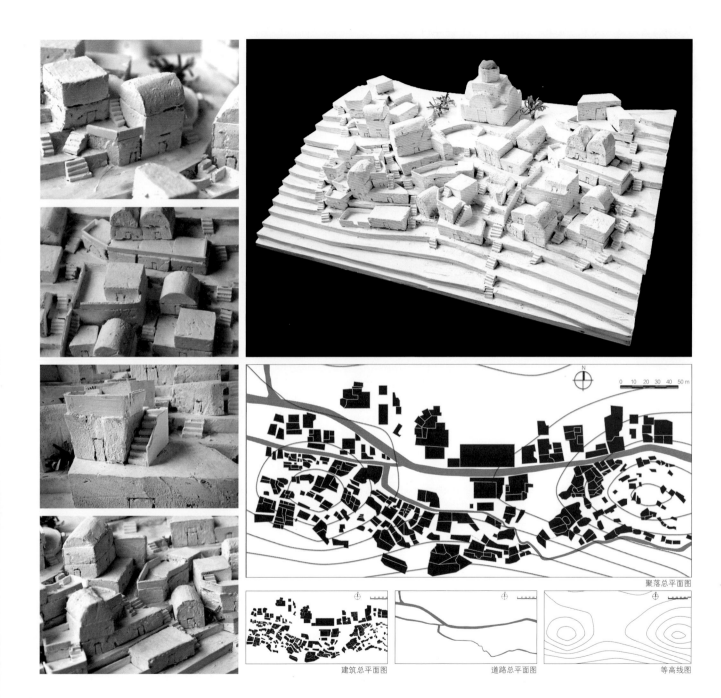

聚落总平面图

建筑总平面图　　道路总平面图　　等高线图

建筑设计基础
ARCHITECTURAL DESIGN FOUNDATION

刘铨　史文娟
LIU Quan, SHI Wenjuan

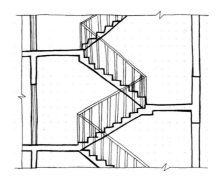

教案设计的背景

"建筑设计基础"课程的主要任务是在中国建筑教育的现实条件下，让原本对建筑学一无所知的新生建立基础性的专业知识架构。其主要内容是建筑认知和建筑表达。认知是主线，表达是方法。认知成果须通过表达方式得以检验，而表达的效果和认知成果直接对应。以这一目标为出发点，本教案提出了建筑设计的基础教学体系，将基础教学分解为针对建筑对象的三个基本建筑关键点，即空间与尺度、结构与构造、场地与环境。

本课程为专业核心课程，教学对象为本科二年级，课程人数为30人。

教案架构

教案的基本架构是在重新认识建筑基础知识的前提下，将认知与表达作为这门课的教学主线，依照循序渐进的原则，分三个阶段设置了不同的教学任务，每个阶段有其特定的认知对象和认知方法，包含若干练习。同时每个阶段的训练都建立在之前一个阶段学习要点的基础上，力图更好地使学生通过认知的过程从一个外行逐步进入专业领域，并为后续的建筑设计学习打下宽阔和扎实的基础。

阶段一：
建筑立面局部测绘：从材料与构件尺寸认知到正投影图绘制；
建筑物测绘：从建筑空间分割与功能尺度认知到平、剖面图绘制；
建筑窗测绘：从建筑构造认知到大样图绘制。
阶段二：
建筑结构模型制作：从结构识图到结构模型制作；
墙身模型制作：从大样识图到构造模型制作。

阶段三：
街道空间认知：理解街巷肌理、城市街道空间及其限定与功能；
地块与建筑类型认知：理解地块肌理、城市建筑类型及其功能与交通组织；
地形与气候认知：理解自然地形、植被及日照等自然环境要素。

教学方法

线上线下结合：本课程使用了慕课作为线上教学工具，方便学生对知识的学习，课堂则解决应用中的具体问题和进行实操指导，同时尽量利用网络分发资料和提交作业。
课堂课外结合：测绘、结构构造认知、城市空间认知阶段都设置了现场讲解，使学生直接面对学习对象，提高了教学效果。

Background of teaching plan design

The main task of the course "Architectural Design Foundation" is to establish a basic professional knowledge framework for freshmen who originally know nothing about architecture under the realistic conditions of Chinese architectural education. Its main content includes architectural cognition and architectural expression. Cognition is the main line and expression is the method. Cognitive achievements need to be tested by means of expression, and the effect of expression corresponds directly to cognitive achievements. Taking this goal as the starting point, this teaching plan puts forward the basic teaching system of

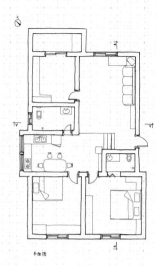 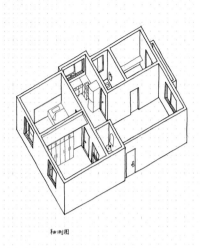 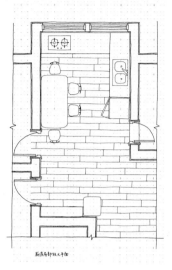

architectural design, which divides the basic teaching into three basic architectural key points for architectural objects, namely space and the scale, structure and construction, the site and environment.
This course is a major core curriculum which contains 30 students in the second year of undergraduate studies.

Teaching plan structure
The basic structure of the teaching plan is to take cognition and expression as the main teaching line of this course under the premise of reacquainting the basic knowledge of architecture. According to the principle of step-by-step, different teaching tasks are set in three stages. Each stage has its specific cognitive objects and cognitive methods, including several exercises. At the same time, the training of each stage is based on the learning points of the previous stage, trying to better enable students to gradually enter the professional field from a layman through the cognitive process, and lay a broad and solid foundation for subsequent architectural design learning.
Stage 1:
Local surveying and mapping of building facade: From recognizing the size of materials and components to drawing orthographic projection;
Building surveying and mapping: From building space segmentation and the functional scale to the plan and section drawings;
Building windows' surveying and mapping: From building structure cognition to detail drawing.

Stage 2:
Building structural model making: From structural drawing recognition to structural model making;
Wall body model making: From detail drawing recognition to structural model making.
Stage 3:
Street space cognition: Understand street texture, urban street space and its limitations and functions;
Cognition of the plot and building type: Understand plot texture, urban building types, and the functions and traffic organization;
Terrain and climate cognition: Understanding natural environment elements such as natural terrain, vegetation and sunlight.

Teaching methods
Combination of online and offline: This course uses MOOC as an online teaching tool to facilitate students' learning. Classes solve specific application questions and gives practical guidance. At the same time, it is useful to make full use of the network to solve data distribution and homework submission.
Combination of classes and extracurriculum: On-site explanation is set for all three phases of surveying and mapping, structural cognition and urban space cognition. Students face the learning objects directly, which improves the teaching effect.

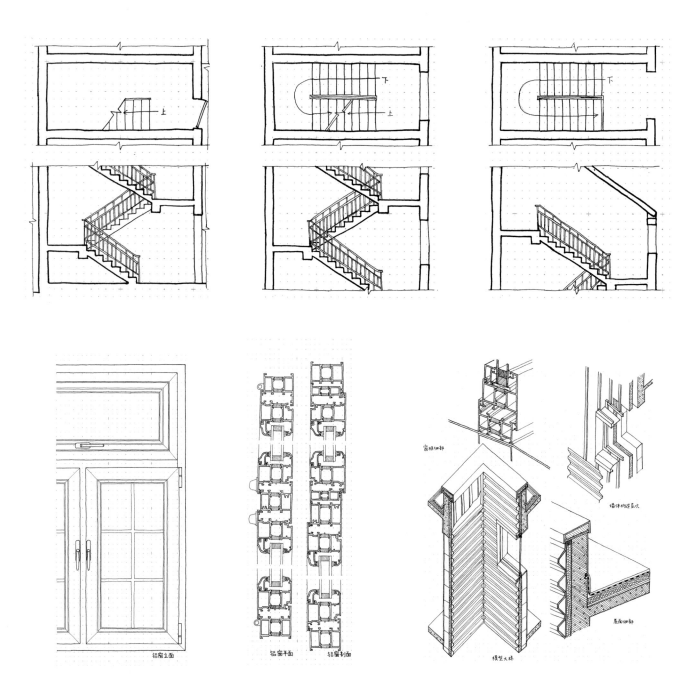

学生：张伊儿 Student: ZHANG Yi'er

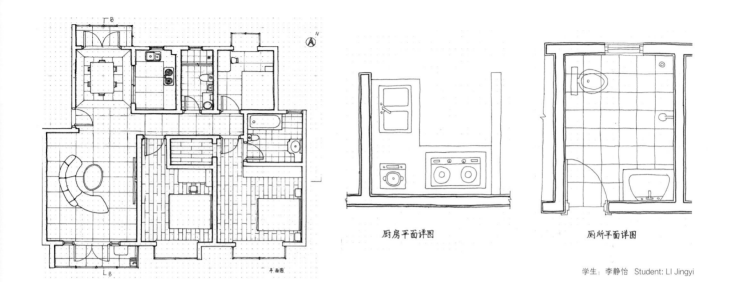

厨房平面详图　　　　厕所平面详图

学生：李静怡　Student: LI Jingyi

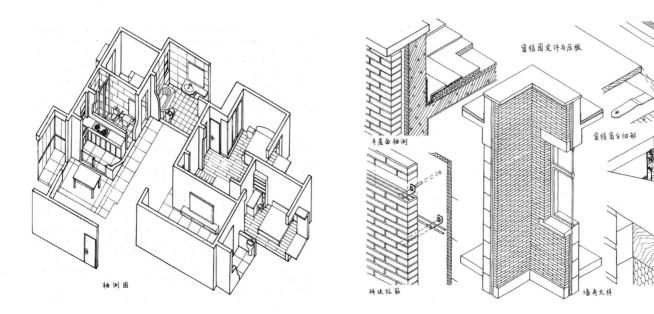

轴侧图

学生：陆星宇　Student: LU Xingyu　　　　　　学生：陆星宇　Student: LU Xingyu

建筑设计（一） ARCHITECTURAL DESIGN 1

限定与尺度：独立居住空间设计
LIMITATIONS AND SCALES: INDEPENDENT LIVING SPACE DESIGN

刘铨　冷天　吴佳维
LIU Quan, LENG Tian, WU Jiawei

教学目标
本次练习的主要任务是综合运用前期案例学习中的知识点——建筑在水平方向上如何利用高度、开洞等操作划分空间，内部空间的功能流线组织及视线关系、墙身、节点、包裹体系、框架结构的构造方式，周围环境对空间、功能、包裹体系的影响等，初步体验一个小型独立居住空间的设计过程。

基本任务
1. 场地与界面：本次设计的场地面积为 80~100 m^2，场地单面或相邻两面临街，周边为 1~2 层的传统民居。
2. 功能与空间：本次设计的建筑功能为小型家庭独立式住宅（附设有书房功能）。家庭主要成员包括一对年轻夫妇和 1~2 位儿童（7 岁左右）。新建建筑面积 160~200 m^2，建筑高度 ≤ 9 m（不设地下空间）。设计者根据设定的家庭成员的职业及兴趣爱好确定空间的功能（职业可以是但不局限于理、工、医、法的技术人员）。
3. 流线组织与出入口设置：考虑建筑内部流线合理性以及建筑出入口与场地周边环境条件的合理衔接。
4. 尺度与感知：建筑中的各功能空间的尺寸需要以人体尺度及人的行为方式作为基本的参照，并通过图示表达空间构成要素与人的空间体验之间的关系。

教学进度
本次设计课程共 6 周。
第一周：构思并撰写几个有代表性的生活场景（家庭人物构成、人物相对关系、类似戏剧中的"折"之剧本）。四个地块分别制作 1:100（60 cm×60 cm）的场地模型。利用 1:100 手绘图纸及 1:100 体块模型构思内部空间及其关系。
第二周：用 1:50 手绘平、立、剖面图纸，结合 1:50 工作模型辅助设计，在初步方案的基础上考虑功能与空间、流线与尺度。
第三周：确定设计方案，制作体现功能关系的空间关系模型（例如立体泡泡图）。
推进剖、立面设计。
第四周：设计深化，细化推敲各设计细节，并建模研究内部空间效果（集中挂图点评）。
第五周：制作 1:20 或 1:30 剖透视图和各分析图，制作 1:50 表现模型。
第六周：整理图纸、排版并完成课程答辩。

成果要求
A1 灰度图纸 2 张，纸质表现模型 1 个（1:50），场地模型 1 个（1:100），工作模型若干。图纸内容应包括：
1. 总平面图（1:200），各层平面图、纵横剖面图和主要立面图（1:50），墙身大样（1:10），内部空间组织剖透视图 1 张（1:20）。
2. 设计说明和主要技术经济指标（用地面积、建筑面积、容积率、建筑密度）。
3. 表达设计意图和设计过程的分析图（体块生成、功能分析、流线分析、结构体系等）。
4. 纸质模型照片与电脑效果图、照片拼贴等。

Teaching objectives
The main task of this exercise is to comprehensively use the knowledge points in the early case study—how to use height, opening and other operations to divide space in the horizontal direction of the building, functional streamline organization and line of sight relationship of internal space, construction mode of the wall body, nodes, wrapping system and frame structure, the influence of the surrounding environment on the space, function and wrapping system, etc. and preliminarily experience the design process of a small independent living space.

Basic tasks
1. Sites and interfaces: The site of this design covers an area of about 80–100 m², facing the street on one side or two adjacent sides, surrounded by 1–2 floors of traditional residential buildings.
2. Functions and space: The building function of this design is a small family independent residence (with study function attached). The main members of the family include a young couple and 1–2 children (about 7 years old). The new-built area is 160–200 m² and the building height is less than 9 m (no underground space). The designer determines the function of the space according to the set occupations and interests of family members (the occupations can be but not limited to technicians of science, engineering, medicine and law).
3. Streamline organization and entrance and exit setting: Consider the rationality of the internal streamline of the building and the reasonable connection between the entrance and exit of the building and the surrounding environmental conditions of the site.
4. Scales and perception: The size of each functional space in the building needs to take the human body scale and human behaviors as the basic reference, and express the relationship between spatial constituent elements and human spatial experience through diagrams.

Teaching progress
This design course takes 6 weeks in total:
Week 1: Conceive and write several representative life scenes (composition of family characters, relative relationship of characters, a script similar to "folding" in drama). Make 1:100 (60 cm × 60 cm) site models for the four plots respectively. Use 1:100 hand-painted drawings and the 1:100 block model to construct the internal functional space and its relationship.
Week 2: 1:50 hand-drawn plan, elevation and section drawings, combined with 1:50 working model to assist design, considering the function and space, streamline and scale on the basis of preliminary scheme.
Week 3: Determine the design scheme, make the spatial relationship model reflecting the functional relationship (such as the three-dimensional bubble diagram), and promote the section and elevation design.
Week 4: Deepen the design, refine the design details, and model and study the internal space effect (centralized wall chart comments).
Week 5: Make 1:20 or 1:30 sectional perspective views and analysis diagrams, and make the 1:50 performance model.
Week 6: Organize drawings, typeset and complete course oral defense.

Achievement requirements
Two A1 grayscale drawings, one paper performance model (1:50), one site model (1:100) and several working models. The contents of the drawings shall include:
1. General plan (1:200), the plan of each floor, the vertical and horizontal section and main elevation (1:50), the wall detail (1:10), the sectional perspective view of internal space organization (1:20).
2. Design descriptions and main technical and economic indicators (the land area, building area, plot ratio and building density).
3. Analysis diagrams expressing design intentions and design processes (the block generation, function analysis, streamline analysis, structural system, etc.).
4. Paper model photos, computer renderings, photo collages, etc..

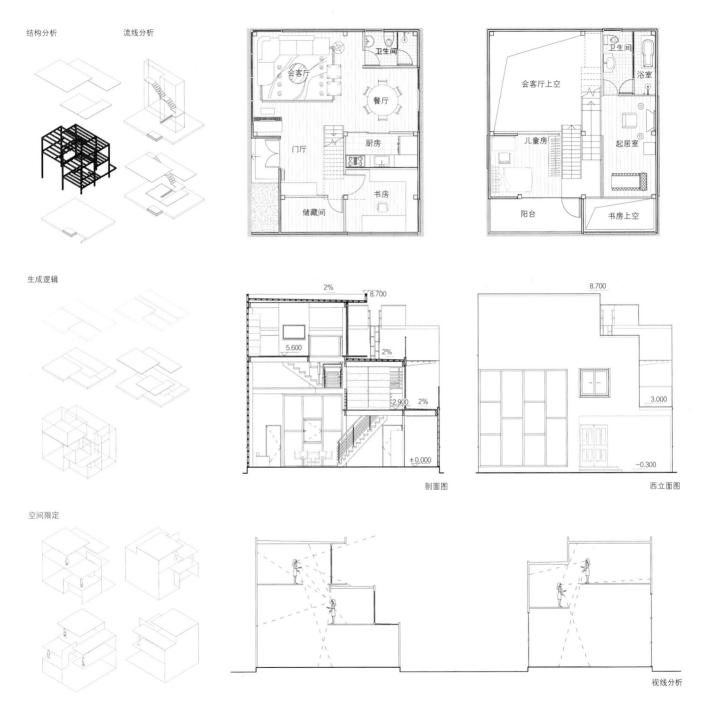

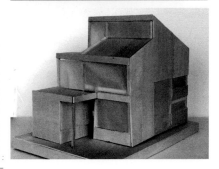

学生：沈至文　Student: SHEN Zhiwen

建筑设计（二）ARCHITECTURAL DESIGN 2
校园多功能快递中心设计
CAMPUS MULTIFUNCTIONAL EXPRESS CENTER DESIGN

刘铨　冷天　吴佳维　孟宪川
LIU Quan, LENG Tian, WU Jiawei, MENG Xianchuan

教学目标
在社会信息化、电商化程度日益提高，疫情常态化的背景下，"快递"活跃且丰富地改变了人们的日常生活，甚至某些非常时期，成为保障基本生活需求的重要方式。其中，高校内的快递行为富集，快递服务渐渐成为了校园后勤服务中不可或缺的一环，与师生日常活动密不可分，成为像食堂、公共浴室、超市一样重要的校园生活区基础公共设施。本次练习的主要任务，是在南京大学鼓楼校区南园建设一个校园多功能快递服务中心，要求综合运用建筑设计基础课程的知识点，操作一个小型公共建筑设计项目。

设计场地
设计训练场地位于南京大学鼓楼校区南园中轴线的西侧界面上，110报警中心所在地块，其南侧是南园的主要建筑教学楼，东北侧是中轴线上的圆形广场，西侧为学生宿舍区。设计范围为地块南半部分，面积大约为701.58 m²左右，新建建筑红线在设计范围的北侧，面积447.96 m²，包括了原有快递服务中心的一层建筑及其西侧的辅助用房。

设计要求
1. 功能流线与活动
公共空间的功能——空间的使用方式——是公共建筑设计的重要内容。对于"快递服务中心"而言，核心功能是快递服务。功能优化方面，作为校园的基础服务设施，可以思考在其中加入补充的特色功能，使快递中心功能更加丰富、复合，帮助提升校园空间体验；或是在疫情常态化的背景下，设置可变空间，增加校园内的弹性空间。基于对场地和功能的思考拟定一份设定计划书，内容包括对快递服务中心的定位思考、快递中心服务性功能的种类、各功能面积配比等。

与食堂、公共浴室类似，快递服务也有使用高峰时段。功能流线的组织对快递空间的使用秩序至关重要，如何合理组织快递上架、师生取寄件等行为流线，是评价快递服务中心空间的重要标准之一。除此之外，还要考虑快递服务中心在校园中的所处位置，通过调研南园现有快递点师生寄取件情况，预设应对快递服务活动对现有校园环境带来的动线、聚集与疏散、多元化社交等系列问题。

2. 形体与环境对话
场地内部与周边的现状建筑与形式特征是塑造新建建筑体量的基本条件。本次设计场地是南京大学鼓楼校区南园内的真实地块。基地处于从校门口到教育超市、公共浴室的必经之路上，交通方便。作为场所环境的实体存在，建筑需要呼应周边的场地环境，充分考虑周围环境要素位置的视线关系。建筑高度≤4.8 m（檐口高度，不包括女儿墙），不超过两层（一层为主，可设置夹层）。新建建筑面积为250~300 m²（±10%）。考虑到建筑位于大学校园内，形体上一方面应具有标识性和整洁性，以丰富校园景观界面；另一方面应该符合校园气质，使建筑具有在地性。利用工作模型辅助设计，探讨建筑形体与功能之间的整合关系，思考如何利用水平、垂直构件来组织空间流线和限定功能，营造丰富的视觉和空间体验。

3. 材料与构造
建筑实体部分最终限定并定义了空间，建构策略可以直接指向形体的生成以及最终效果。选择合适的材料和结构形式，利用建构的思维提出解决空间问题的策略，呼应场地需求。

教学进度
第一周：场地认知，结合已有图像资料对快递服务中心的建设场地进行实地调研，包括场地现状及周边交通和环境信息。案例分析及相关资料收集，了解快递服务中心的运营方式、运作流线、空间特征及操作思路和具体策略。

第二周：制作场地模型（底座60 cm×60 cm×5 cm），比例1：100。对场地有初步了解之后，对场地现有问题做出批判并提出优化方向，以此作为依据生成设计概念。自拟快递驿站复合的新功能，写出详细的设定计划书，包括快递中心服务性功能的种类、各功能面积配比等，确定初步概念和设计方向。

第三周：思考与场地周围现有建筑的对话关系，提出处理设计问题的结构策略，解决使用流线问题，整合结构与空间的组织方式。形成初步方案与工作模型，比例1：80。

第四周：深化方案阶段，优化并发展前述的结构策略，用1：50的图纸比例，手绘平、立、剖面图纸，在初步方案的基础上深化实体与建造层面的思考（中期集体挂图点评）。

第五周：制作工作模型辅助设计，进一步优化结构设计，使得结构部分清晰明确可认知，明确支撑体和围护的各自作用。

第六周：确定最终的设计方案，并将研究的重心转移到建构研究部分（集体挂图并进行模型点评）。

第七周：了解实际建造所面临的误差问题和节点问题，深化1：50图纸，细化推敲各设计细节，制作全新的结构体模型，比例1：50（各小组挂图并点评模型）。

第八周：排版调整，思考并选择图面表达的效果，制作必要的分析图和效果图。整理并完成图纸，制作正式模型（基地模型1：100）并完成课程答辩。

Teaching objectives
With the development of increasing social informatization and e-commerce, and the normalization of the epidemic, "express delivery" has actively and richly changed people's daily lives, and even during some specific periods, it becomes an important way to ensure basic living needs. Among them, express delivery behaviors in colleges and universities are enriched, and express delivery service has gradually become an indispensable part of campus logistics services. It is inseparable from the daily activities of teachers and students, and has become an important basic public service facility in campus living areas like canteens, public bathrooms, and supermarkets. The main task of this exercise is to build a campus multi-functional express service center in the South Park of Gulou Campus of Nanjing University. It is required to comprehensively use the knowledge points of the basic course of architectural design to operate a small public building design project.

Design site
The design training site is located on the west interface of the central axis of the South Park of the Gulou campus of Nanjing University. On the south side of the land where the police centre is located stands the main teaching buildings of the South Park, on the northeast side is the circular square on the central axis, and on

the west side is the student dormitory area. The design scope is the southern half of the plot, with an area of about 701.58 m². The red line of the new building is on the north side of the design scope, with an area of about 447.96 m², including the one-story building a's the original express service center and the auxiliary buildings on the west side.

Design requirements
1. Functional flow and activities
The function of public space and the way that space is used are important aspects of the design of public buildings. For "express service center", the core function is express service. In terms of function optimization, as the basic service facilities of the campus, we can consider adding supplementary features to make the functions of the express delivery center richer and more complex, and help improve the campus space experience; or in the context of the normalization of the epidemic, setting up a variable space can increase the flexible space on campus. Based on the consideration of the site and functions, a setting plan is drawn up, which includes thinking about the positioning of the express service center, the types of service functions of the express center, and the proportion of each functional area, etc..
Similar to canteens and public bathrooms, express service center also has peak hours of use. The organization of the functional streamline is crucial to the use order of the express delivery space. How to rationally organize the streamline of express deliveries on the shelves, teachers and students picking up deliveries, etc. is one of the important criteria for evaluating the express service center space. In addition, the location of the express service center in the campus should also be considered. By investigating the situation of teachers and students sending and picking up items at the existing courier points in South Park, it is necessary to presuppose the solutions to problems such as streamlines, gather and evacuation, diversified social interaction and other issues brought by express service activities on the existing campus environment.
2. Interaction between forms and environments
The existing architecture and form characteristics in and around the site are the basic conditions for shaping the volume of the new building. The site for this design is a real plot in the South Park of the Gulou Campus of Nanjing University. The base is on the only way from the school gate to the supermarket and public bathrooms, with convenient transportation. As the physical existence of the site environment, the building needs to respond to the surrounding environment and fully consider the line of sight relationship between the positions of the surrounding environmental elements. The building height is less than 4.8 m (cornice height, excluding parapet), no more than two floors (mainly one floor,

interlayer can be set). The new construction area is about 250–300 m² (±10%). Considering that the building is located on the university campus, on the one hand, its shape should be noticeble and tidy to enrich the campus landscape interface; on the other hand, it should conform to the campus features, making the building local. Use the working model to assist design, explore the integration relationship between architectural forms and functions, think about how to use horizontal and vertical components to organize space streamlines and limit functions, and create a rich visual and spatial experience.
3. Materials and construction
The physical part of the building finally limits and defines the space, and the construction strategy can directly point to the generation of the shape and the final effect. Choose appropriate materials and structural forms, use constructive thinking to propose strategies to solve space problems, and respond to site needs.

Teaching progress
Week 1: Site awareness, combined with the existing image data to conduct field research on the construction site of the express service center, including the current situation of the site and surrounding traffic and environmental information. Do some case analysis and related data collection to understand the operation mode, operation streamline, space characteristics, operation ideas and specific strategies of the express service center.
Week 2: Make a 1∶100 site model (base 60 cm × 60 cm × 5 cm). After a preliminary understanding of the site, criticize the existing problems of the site and propose an optimization direction, and use this as a basis to generate a design concept. Design new functions of the express service center, write a detailed setting plan, including the types of service functions of the express center, the proportion of each functional area, etc. Determine the preliminary concept and design direction.
Week 3: Think about the dialogue relationship with the existing buildings around the site, propose structural strategies to deal with design issues, solve the problem of using streamlines, and integrate the organization of structures and space. Form a preliminary plan and a working model (1∶100).
Week 4: Deepen the plan, optimize and develop the aforementioned structural strategy, use the drawing scale of 1∶50, draw plan, evelation and section by hand, and deepen the thinking on the entity and construction level on the basis of the preliminary plan (mid-term collective wall chart comment).
Week 5: Make a working model to aid in the design, further optimize the structural design, make the structural part clear and recognizable, and clarify the respective roles of the support body and the enclosure.
Week 6: Determine the final design scheme, and shift the focus of the research to the construction research part (collective wall chart and model review).
Week 7: Understand the error and node problems faced by the actual construction, deepen the 1∶50 drawings, refine the design details, and make a new structural model with a scale of 1∶50 (each group wall chart and comment on the model).
Week 8: Typesetting adjustment, think and choose the effect of graphic expression, make necessary analysis diagrams and effect diagrams. Organize and complete the drawings, make a formal model (base model 1∶100) and complete the course oral defense.

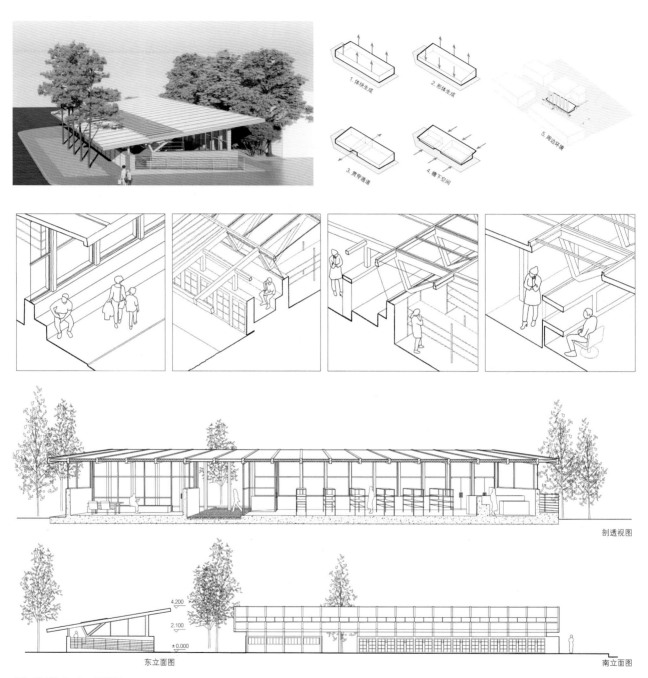

剖透视图

东立面图   南立面图

学生：陈浏毓 Student: CHEN Liuyu

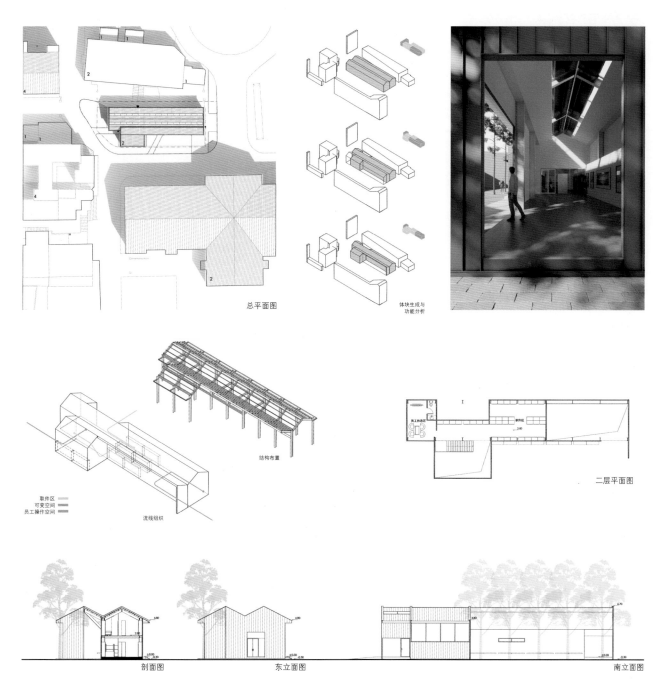

建筑设计（三）ARCHITECTURAL DESIGN 3

专家公寓设计
THE EXPERT APARTMENT DESIGN

童滋雨　窦平平　黄华青
TONG Ziyu, DOU Pingping, HUANG Huaqing

教学目标

从空间单元到系统的设计训练。

从个体到整体，从单元到体系，是建筑空间组织的一种基本和常用方式。本课题首先关注空间单元的生成，并进一步根据内在的使用逻辑和外在的场地条件，将多个单元通过特定方式与秩序组合起来，形成一个兼备合理性、清晰性和丰富性的整体系统。基本单元的重复、韵律、变异等都是常用的操作手法。

基本任务

拟在南京大学鼓楼校区南园宿舍区内新建专家公寓1座，用于国内外专家到访南京大学开展学术交流活动期间的居住。用地位于南园中心喷泉西侧，面积约3 600 m²。地块上原有建筑将被拆除，新建建筑总建筑面积不超过3 000 m²。高度不超过3层。具体的功能空间包括：

客房：30间左右，分为单间和套间两类，单间面积为35~40 m²，套间面积为70~80 m²，内部需包括睡眠空间、卫生间、学习空间、工作空间。套间可考虑必要的接待空间和简单的餐厨空间。套间不少于5间。

大会议室1间：100~120 m²。

研讨室3间：每间约60 m²。

休闲区与咖啡吧（兼做早餐厅）：约150 m²。

操作间：约30 m²。

服务间：每层1间，每间约20 m²。

工作人员办公室与休息室：1~2间，每间约30 m²。

其他必要的门厅、前台、公共卫生间、储藏室、服务间等自行设置。

场地环境：结合建筑总体布局，在建筑周边及其内部创造优美的室内外场地环境，供使用者休憩交往，并为校园增色。

成果要求

图纸：总平面图（1:500），建筑平、立、剖面图（1:200），客房单元平面图（1:50），分析图、轴测图、鸟瞰图、剖透视图、人眼透视图，其他有助于表达方案的图纸。

手工模型：比例1:200。

教学进度

阶段一：场地调研分析，背景与案例研究，空间单元体的生成（2周）；

阶段二：单元的组合与总体布局（1周）；

阶段三：设计深化（3周）；

阶段四：最后成果制图、排版，准备答辩（2周）。

Teaching objectives

From space units to systematic design training.

From individuals to the whole, from units to the system, it is a basic and common way of architectural space organization. This topic first pays attention to the generation of spatial units, and further combines multiple units with order in a specific way according to the internal use logic and external site conditions to

form an overall system with rationality, clarity and richness. Repetition, rhythm and variation of basic units are commonly used.

Basic tasks
It is proposed to build a new expert apartment in the South Park dormitory area of Gulou campus of Nanjing University for domestic and foreign experts to live during their visit to Nanjing University for academic exchange activities. The land is located in the west of the fountain in the center of South Park, covering an area of about 3 600 m². The original buildings on the plot will be demolished, and the total construction area of the new building will not exceed 3 000 m². The height shall not exceed 3 floors. Specific functional spaces include:
Guest rooms: about 30 rooms, divided into single rooms and suites, with a single room area of 35–40 m² and a suite area of 70–80 m². The interior spaces need to include the sleeping space, bathroom, study space and work space. The necessary reception and simple kitchen space can be considered in the suite. No less than 5 suites.
1 large conference room: 100–120 m².
3 seminar rooms: Each about 60 m².
Leisure area and the coffee bar (also as a breakfast restaurant): About 150 m².
The operation room: About 30 m²
The service room: One room on each floor, about 20 m² each.
Staff offices and lounges: 1–2, each about 30 m².
Other necessary foyers, front desks, public toilets, storage rooms, service rooms, etc. shall be set accordingly.
Site environment: Combine with the overall layout of the building, create a beautiful indoor and outdoor site environments around and inside the building for users to rest and communicate, and enhance the landscape of the campus.

Achievement requirements
Drawings: The general plan drawing (1:500), building plan, elevation and section drawings (1:200), the guest room unit plan (1:50), analysis drawings, axonometric drawings, aerial views, sectional perspective views, human eye perspective views, and other drawings which are helpful to express the scheme.
In manual model: Scale 1:200.

Teaching progress
Stage 1: Site investigation and analysis, background and case study, generation of spatial unit (2 weeks);
Stage 2: Unit combination and overall layout (1 week);
Stage 3: Design deepening (3 weeks);
Stage 4: Final achievement drawing, typesetting and preparation of oral defense (2 weeks).

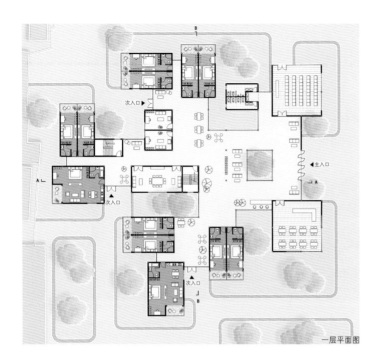
一层平面图

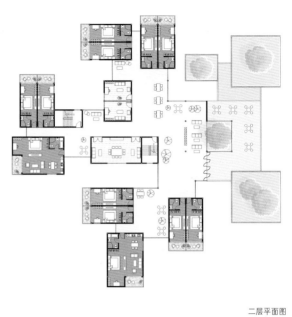
二层平面图

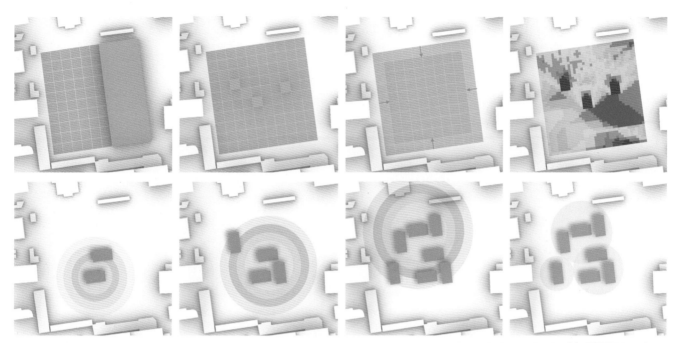

学生：夏月 Student: XIA Yue

建筑设计(四) ARCHITECTURAL DESIGN 4

世界文学客厅
WORLD LITERATURE LIVING ROOM

华晓宁　窦平平　黄华青
HUA Xiaoning, DOU Pingping, HUANG Huaqing

教学目标

本课程主题是"空间",学生学习建筑空间组织的技巧和方法,训练对空间的操作与表达。空间是建筑学的基本。本课题基于文学主题,训练文本、叙事与空间序列的串联,学生学习空间叙事与空间用途的整体构思,充分考虑人在空间中的行为、空间感受,尝试以空间为手段表达特定的意义和氛围,最终形成一个完整的设计。

设计场地

南京古称金陵、白下、建康、建邺……历来是人文荟萃、名家辈出之地,号称"天下文枢"。南京作为六朝古都,亦为中国文学之始。何为文？梁元帝曰："吟咏风谣,流连哀思者,谓之文。"汉魏有文无学,六朝文学《文选》《文心雕龙》《诗品》既是文学评论的开始,也是文学的发端。

2019年,南京入选联合国"世界文学之都",开展一系列城市空间计划,包括筹建"世界文学客厅",作为一座以文学为主题的综合性博物馆。该馆选址位于北极阁公园东南隅,用地面积约为5 050 m²,紧临市政府中轴线,毗邻古鸡鸣寺、玄武湖、明城墙、东南大学四牌楼校区等历史文化遗迹,构成城市与山林之间的过渡空间。设计应妥善处理建筑与周边城市环境和既有建筑的关系,彰显中国文学的精神特质。

空间计划

建筑总高度不超过12 m,容积率不超过0.5,绿地率不低于25%

空间组织要有明确特征和明确意图,概念清晰;满足功能合理、环境协调、流线便捷的要求。注意不同类型和不同形态空间的构成、空间的串联组织和空间氛围的塑造。

总建筑面积:约为2 500 m²。

(以下各部分面积配比为参考,每人可以根据研究自行策划并进行适当调整。)

1. 世界文学之都展示中心

主题展厅:400~500 m²;

国际互联厅:200 m²;

临时展厅:100~150 m²;

其他辅助空间(控制室、储藏间等)。

2. 国际交流中心

报告厅(100座):150~200 m²;

会议室(4~6间):共300 m²;

会客厅:60~80 m²;

其他辅助空间(休闲、接待等)。

3. 城市客厅

游客服务(可结合门厅):150~200 m²。

4. 旧书馆

书籍展示及阅览:150~200 m²。

5. 行政办公与辅助空间

办公室:6间,每间不小于15 m²;

专家接待室:4间,每间不少于30 m²;

其他门厅、交通、设备间、卫生间等面积根据设计需要自行确定。

6. 场地

园林景观、户外展场、停车场等,应考虑建筑与景观的整体关系,以景观烘托氛围。

Teaching objectives

The theme of this course is "space", students will learning the skills and methods of architectural space organization, and train the operation and expression of space. Space is the base of architecture. Based on the literary theme, this topic trains the series connection of the text, narration and spatial sequence. Students learn the overall idea of spatial narration and spatial use, fully considers people's behaviors and the spatial feelings in space, try to express specific meanings and atmosphere by means of space, and finally form the complete design.

Design site

In ancient times, Nanjing was called Jinling, Baixia, Jiankang and Jianye…It has always been a place with a large number of talents and famous scholars, known as the "cultural hub of the world". As the ancient capital for the Six Dynasties, Nanjing is also the beginning of Chinese literature. What is literature? Emperor Liang Yuandi said, "It is called literature to chant wind ballads and linger around mourning." The Han and Wei Dynasties have articles but no literature. The literary works of the Six Dynasties such as *Selected Works*, *Literary Heart and Carving Dragons* and *Poetry* is not only the beginning of literary criticism, but also the beginning of literature.

In 2019, Nanjing was selected as the "capital of world literature" by the United Nations and planned to carry out a series of urban space plans, including preparing to build the "world literature living room" as a comprehensive museum with literature as the theme. The museum is located in the southeast corner of Beijige Park, with a land area of about 5 050 m^2. It is close to the central axis of the municipal government and adjacent to historical and cultural sites such as ancient Jiming Temple, Xuanwu Lake, Ming City Wall and Sipailou campus of Southeast University, forming a transition space between the city and mountains. The design should properly deal with the relationship between the building and the surrounding urban environment and existing buildings, and highlight the spiritual characteristics of Chinese literature.

Space plan

The total height of the building shall not exceed 12 m, the plot ratio shall not exceed 0.5, and the green space rate shall not be less than 25%.

Spatial organization should have clear characteristics, clear intentions and clear concepts; meet the requirements of reasonable functions, coordinated environment and convenient streamlines. Pay attention to the composition of different types and forms of space, the series connection organization of space and the shaping of space atmosphere.

Total construction area: About 2 500 m^2.

(The area ratios of the following parts are for reference, and each person can make appropriate adjustment according to the research.)

1. Exhibition center of the Capital of World Literature

Theme exhibition hall: 400–500 m^2;

International interconnection exhibition hall: 200 m^2;

Temporary exhibition hall: 100–150 m^2;

Other auxiliary spaces (control room, storage room, etc.).

2. International exchange center

Lecture hall (100 seats): 150–200 m^2;

Meeting room (4–6 rooms): 300 m^2 in total;

Reception rooms: 60–80 m^2;

Other auxiliary spaces (leisure, reception, etc.).

3. City living room

Tourist service (can be combined with the lobby): 150–200 m^2.

4. Old library

Book display and reading: 150–200 m^2.

5. Administrative office and auxiliary space

Office: 6 rooms, each not less than 15 m^2;

Expert reception room: 4 rooms, each not less than 30 m^2;

The area of other foyers, traffic, equipment rooms and toilets shall be determined according to the design needs.

6. Site

The overall relationship between architecture and landscapes shall be considered for the garden landscape, outdoor exhibition hall and parking lot, etc., so as to set off the atmosphere with landscapes.

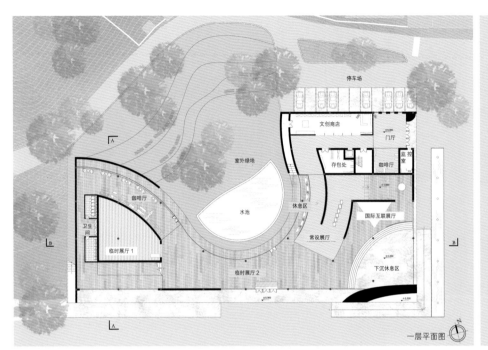

一层平面图

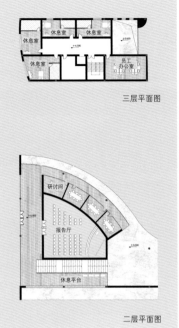

三层平面图

二层平面图

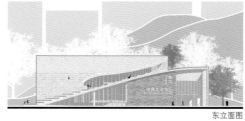
东立面图

南立面图

剖面图 A-A

剖面图 B-B

建筑设计（五）ARCHITECTURAL DESIGN 5

大学生健身中心改扩建设计
RECONSTRUCTION AND EXPANSION DESIGN OF THE COLLEGE STUDENT FITNESS CENTER

傅筱　钟华颖　孟宪川
FU Xiao, ZHONG Huaying, MENG Xianchuan

教学内容

本学期建筑设计课程训练的主题是"城市建筑"。城市，是建筑与建筑师最重要的舞台。从物质环境上来看，城市是由多样化的建筑实体及其之间的空间共同构成的，承载和容纳了城市居民纷繁复杂的生活，以及城市中各种物质、能量、信息的流动和变化。当代复杂多元的城市生活产生了错综复杂的城市物质空间系统，它赋予城市中的建筑诸多限定。而城市中的建筑一旦形成，也即介入和重构了城市物质空间环境，乃至改变和重新定义城市生活本身。在某种程度上，"'事物之间'的形式比事物本身的形式更重要"。这使得建筑设计不再仅仅停留于单纯的自我关注和自我完善，而必须对于所在的城市环境做出积极的应对。另一方面，当代城市生活往往要求更大的建筑规模和更为混杂而灵活的功能混合，随之而来的是建筑自身的高度复杂性。当代建筑师所面临的许多任务日益成为一个庞大的机能运作系统，建筑师在处理和协调诸多规范、技术条件限制上所花的精力往往要超过花在所谓"创意"上的精力。如何使得这些限制对于设计发挥积极的推动作用，是当代建筑师所面临的一个主要挑战。设计的驱动力同时来自两方面：外在的城市运作系统和内在的建筑运作系统，这两个系统的无缝对接、互动和融合是我们所追求的目标。

教学目标

本课题针对复杂城市环境中复杂功能建筑的流线处理、功能与空间组织、外部形态控制，希望在"外""内"两方面达成以下的目标：

1. 分析较为复杂的城市环境，对外部限定条件做出积极合理的回应；
2. 理解城市建筑实体与城市空间的相互界定，将建筑实体与空间融入城市物质空间系统；
3. 初步了解城市公共空间的空间行为特征，学习创造积极而富有活力的公共空间；
4. 学习使用平面化和立体化的多种方法，合理组织建筑内外不同类型的流线；
5. 学习组织建筑内部较为复杂的功能布局与空间分化；
6. 进一步训练建筑造型手法；
7. 初步掌握公共建筑常用的设计规范。

Teaching contents

The theme of this semester's architectural design course training is "urban architecture". Cities are the most important stage for architecture and architects. From the perspective of material environment, the city is composed of diverse architectural entities and the spaces between them, which carry and accommodate the complicated life of urban residents, as well as the flow and change of various substances, energy and information in the city. The contemporary complex and pluralistic urban life has produced a complex urban

physical space system, which gives many limits to the buildings in the city. Once the building in the city is formed, it intervenes in and reconstructs the physical space environment of the city, and even changes and redefines the urban life itself. In a way, "the form 'between things' is more important than the form of things themselves." This makes the architectural design no longer just stay in the simple self-concern and self-improvement, but must make a positive response to the urban environment. Contemporary urban life, on the other hand, often demands a larger architectural scale and a more flexible mix of functions, with this comes a high degree of complexity in the building itself. Many of the tasks that contemporary architects face have increasingly become the construction of a large functioning system, and architects often spend more energy in dealing with and coordinating the many normative and technical constraints than they spend on so-called "creativity". How to make these constraints play a positive role in the design is a major challenge faced by contemporary architects. The driving force of the design comes from two aspects at the same time: the external urban operation system and the internal architectural operation system, and the seamless connection, interaction and integration of the two systems is our goal.

Teaching objectives
This topic aims at the streamline processing, function and space organization, and external form control of complex functional buildings in complex urban environment, hoping to achieve the following goals in both "outside" and "inside" aspects:
1. Analyze the complex urban environment and make positive and reasonable responses to external constraints;
2. Understand the mutual definition of urban architectural entities and urban space, and integrate architectural entities and space into urban physical space system;
3. Preliminarily understand the spatial behavior characteristics of urban public space, and learn to create active and dynamic public space;
4. Learn to use various methods of planarization and trivialization to reasonably organize different types of circulation inside and outside the building;
5. Learn to organize complex functional layouts and spatial differentiation inside buildings;
6. Further train architectural modeling techniques;
7. Master the common design norms of public buildings.

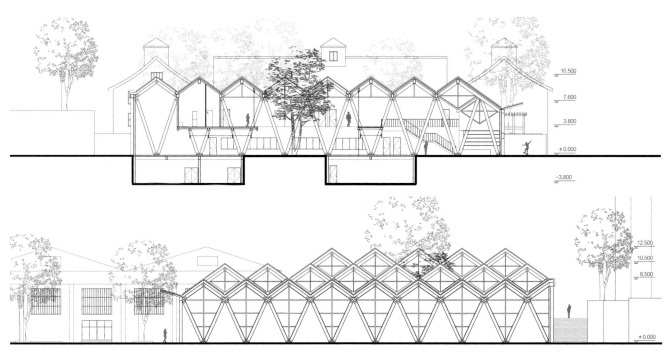

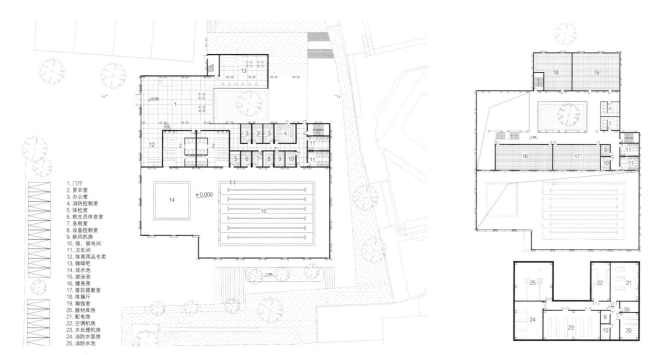

1. 门厅
2. 更衣室
3. 办公室
4. 消防控制室
5. 体检室
6. 救生员休息室
7. 急救室
8. 设备控制室
9. 新风机房
10. 强、弱电间
11. 卫生间
12. 体育用品专卖
13. 咖啡吧
14. 戏水池
15. 游泳池
16. 健身房
17. 普拉提教室
18. 体操厅
19. 瑜伽室
20. 器材库房
21. 配电房
22. 空调机房
23. 水处理机房
24. 消防水泵房
25. 消防水池

学生：黄辰逸　Student: HUANG Chenyi

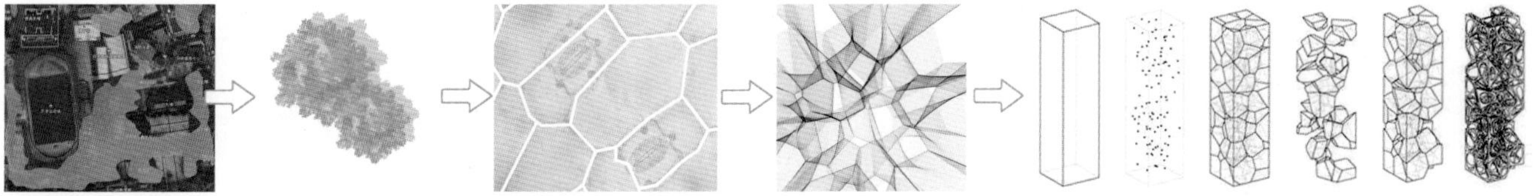

1. 较高的绿植覆盖率　2. 树木、树叶　3. 绿植有机体的生长结构　4.Voronoi 生长结构　5. 建筑结构与覆盖的规则

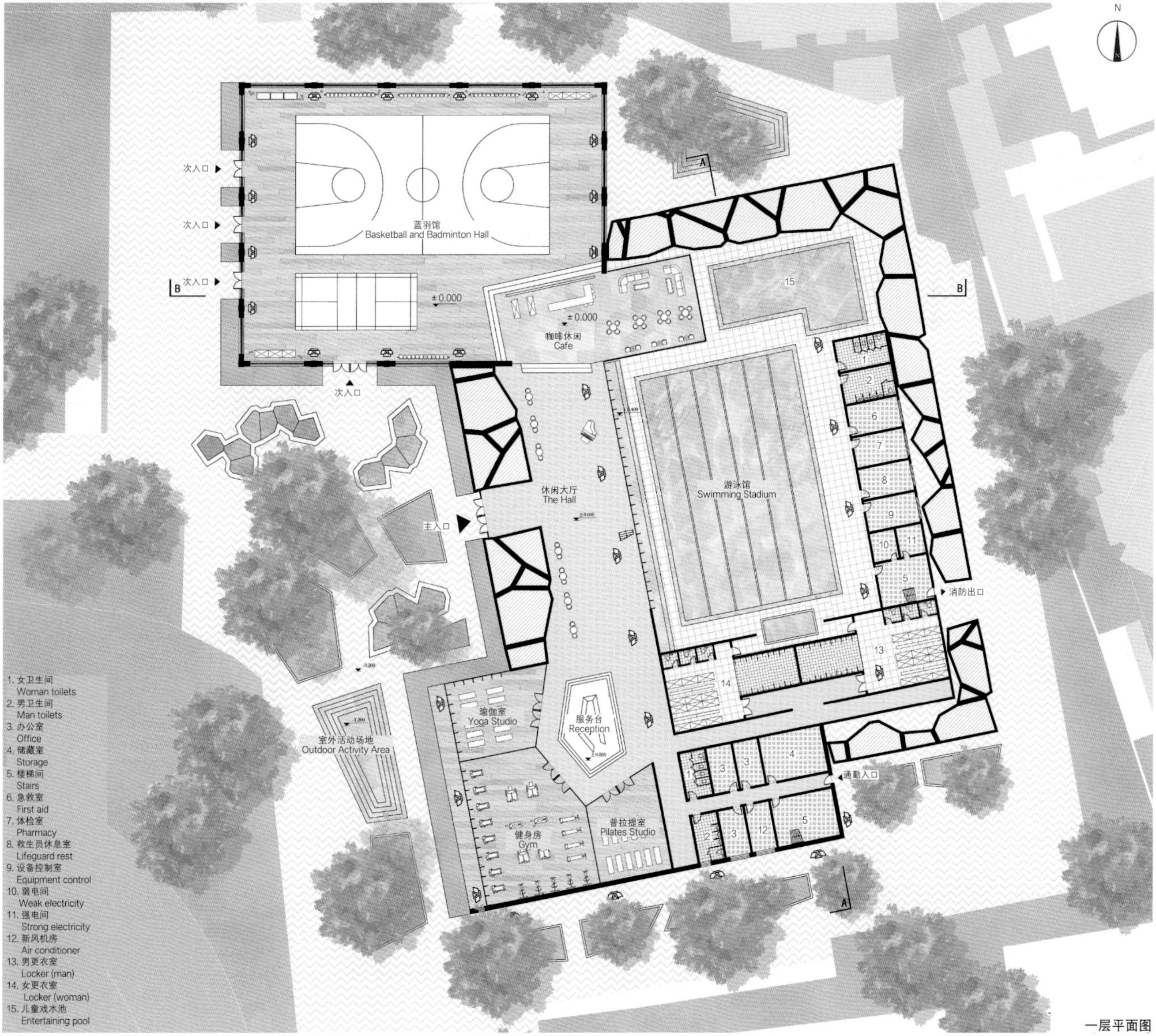

N
蓝羽馆
Basketball and Badminton Hall
次入口
次入口
次入口
次入口
±0.000
咖啡休闲
Cafe
±0.000
15
B
B
A
A
休闲大厅
The Hall
主入口
游泳馆
Swimming Stadium
消防出口
通勤入口
服务台
Reception
瑜伽室
Yoga Studio
健身房
Gym
普拉提室
Pilates Studio
室外活动场地
Outdoor Activity Area
1. 女卫生间
Woman toilets
2. 男卫生间
Man toilets
3. 办公室
Office
4. 储藏室
Storage
5. 楼梯间
Stairs
6. 急救室
First aid
7. 体检室
Pharmacy
8. 救生员休息室
Lifeguard rest
9. 设备控制室
Equipment control
10. 弱电间
Weak electricity
11. 强电间
Strong electricity
12. 新风机房
Air conditioner
13. 男更衣室
Locker (man)
14. 女更衣室
Locker (woman)
15. 儿童戏水池
Entertaining pool

一层平面图

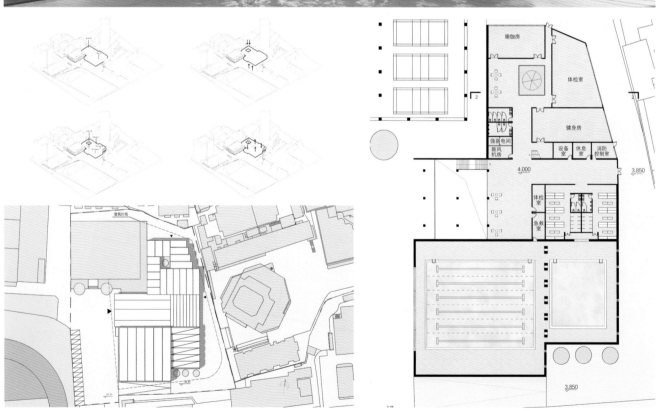

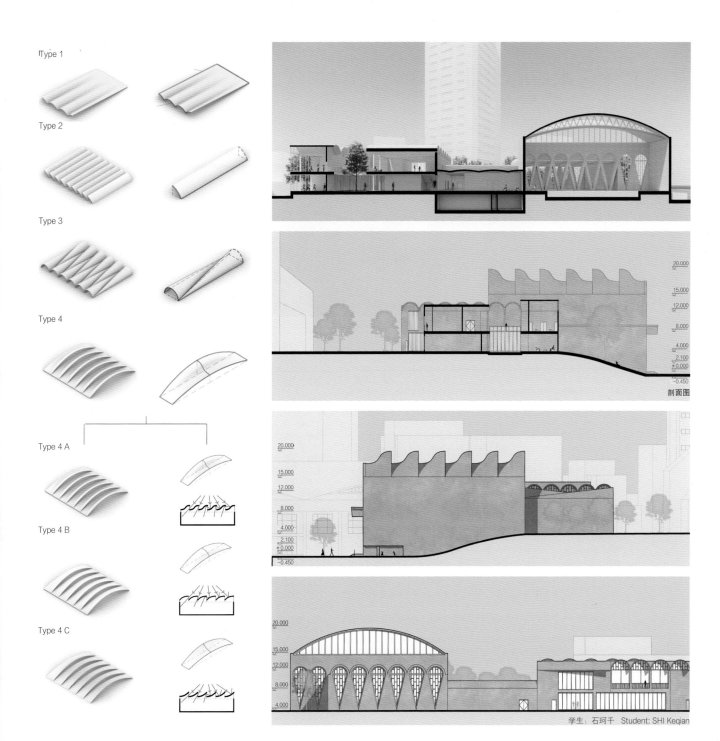

社区文化艺术中心设计
DESIGN OF COMMUNITY CULTURE AND ART CENTER

张雷 王铠 尹航
ZHANG Lei, WANG Kai, YIN Hang

概况
项目拟在百子亭风貌区基地处新建社区文化中心，总建筑面积约为 8 000 m²，项目不仅为周边居民文化基础设施服务，同时也期望成为复兴老城的街区活力的文化地标。根据基地条件、功能使用进行建筑和场地设计。基地用地面积为 4 600 m²。

民国时期，百子亭一带属于高级住宅区，在位置上紧邻作为文教区的鼓楼，以及作为市级行政区的傅厚岗地区。凭借区域上的优势与政府的扶持，百子亭一带自1930年代开始，逐渐成为当时文化精英、社会名流与政府要员的聚居之地。众多受邀前往南京创建其事业的学者、文人都在此购地立业，并建造出了一幢幢"和而不同"的新式住宅。这些建筑既是近代南京城市肌理中的现代图景，也是当时中国有为之士们实践其梦想的舞台，还是中国近现代建筑史中不可忽视的华美段落。

基地条件：现状
根据《南京历史文化名城保护规划（2010—2020）》，百子亭历史风貌区被列入"保护名录"，历史风貌区内现有市级文物保护单位3处，为桂永清公馆旧址、徐悲鸿故居和傅抱石故居，不可移动文物8处，历史建筑1处。

教学内容
1. 演艺中心。包含一个400座小剧场，乙级。台口尺寸 12 m×7 m。根据设计的等级确定前厅、休息厅、观众厅、舞台等面积。观众厅主要为小型话剧及戏剧表演而设置。按60~80人布置化妆室及服装道具室，并设2~4间小化妆室。要求有合理的舞台及后台布置，应设有排练厅、休息室、候场区以及道具存放间等设施，其余根据需要自定。

2. 文化中心。定位于区级综合性文化站，包括公共图书阅览室、电子阅览室、多功能厅、排练厅以及辅导培训、书画创作等功能室（不少于8个且每个功能室面积应不低于 30 m²）。

3. 配套商业。包含社区商业以及小型文创主题商业单元。其中社区商业为不小于 200 m² 超市1处，文创主题商业单元面积为 60~200 m²。

4. 其他。变电间、配电间、空调机房、售票、办公、厕所等服务设施根据相关设计规范确定，各个功能区可单独设置，也可统一考虑。地上不考虑机动车停车配建，街区地下统一解决，但需要根据建筑功能面积计算数量。

教学成果
每人不少于4张A1图纸，图纸内容包括：
1. 城市与环境：总平面图 1：500、总体鸟瞰图、轴测图。
2. 空间基本表达：平、立、剖面图 1：200~1：400。
3. 空间解析与表现：概念分析图、空间构成分析图、轴测分析图、剖透视图（不少于2张，必须包含大空间、公共空间的剖透视图）、室内外人眼透视图若干。
4. 手工模型：每个指导教师组做1个1：500总平面图体量模型，每位学生做1个1：500的概念体块模型。

教学进度
本次设计课程共8周。
第一周：授课（1学时）、调研场地及案例、制作场地模型（SU模型+实体模型 1：500）相应的案例资料收集。
第二周：学生收集案例汇报、初步概念方案讨论（包含体块与场地关系布局、内部空间基本布局）。
第三周：概念深化，完善初步建筑功能布局和空间形态方案（包括基本空间单元及其组合），制作空间形态模型。
第四周：方案定稿，明确空间表皮、平面功能、街区环境模式。
第五周：方案深化Ⅰ，空间表皮、平面功能、街区环境深化。
第六周：方案深化Ⅱ，细化表皮处理、剧场空间及其他重要公共空间的设计。
第七周：方案表达，完成平、立、剖面图绘制，完善SU设计模型。
第八周：制图、排版。

Overview
The project plans to build a new community culture and art center at the base of the Baiziting historical area, with a total construction area of about 8 000 m². The project not only serves as the cultural infrastructure for surrounding residents, but also hopes to become a cultural landmark to revive the vitality of the old town. The building and site design should consider the base conditions and functional use. The land area of the base is 4 600 m².

During the period of the Republic of China, Baiziting was a high-grade residential area, close to the Gulou as the cultural and educational area and the Fuhougang area as the municipal administrative area. With regional advantages and government support, Baiziting area gradually became a settlements for cultural elites, celebrities and government officials since the 1930s. Many scholars invited to Nanjing to establish their careers bought land there and built new houses of "harmony with difference". These buildings are not only the modern landscapes

in the modern urban texture of Nanjing, but also stages for Chinese promising people to practice their dreams at that time, and the gorgeous paragraphs that can not be ignored in the history of Chinese modern architecture.

Base conditions: Current situation

According to the Plan for the *Protection of Famous Historical and Cultural Sites in Nanjing (2010–2020)*, the Baiziting historical area is included in the "protection list". There are 3 municipality protected historic sites in the historical area, including the former site of Gui Yongqing residence, Xu Beihong's former residence and Fu Baoshi's former residence, 8 immovable cultural relics and 1 historical building.

Teaching contents

1. Performing arts center. It contains a small theatre with 400 seats, class B. The size of the proscenium is 12 m × 7 m. Determine the area of the front hall, lounge, auditorium and stage according to the design level. The auditorium is mainly set up for small-scale dramas and drama performance. The dressing room and clothing props room shall be arranged for 60–80 people, and 2–4 small dressing rooms shall be set. The reasonable stage and backstage layout are required. The rehearsal hall, lounge, waiting area, props storage room and other facilities shall be set, and the rest shall be determined according to needs.
2. Cultural center. Serving as the district level comprehensive cultural station, it includes the public reading room, electronic reading room, multi-functional hall, rehearsal hall, counseling and training, calligraphy and painting creation and other functional rooms (no less than 8, and the area of each functional room shall not be less than 30 m^2).
3. Supporting business. It includes community business and small cultural and creative theme business units. Among them, the community business is a supermarket with an area of no less than 200 m^2, and the area of cultural and creative theme business units is 60–200 m^2.
4. Others. Service facilities such as the substation room, power distribution room, air conditioning room, ticket office, office and toilet are determined according to relevant design specifications. Each functional area can be set separately or considered uniformly. The parking allocation of motor vehicles is not considered on the ground, and the underground of the block allocate parking spaces uniformly, but the quantity needs to be calculated according to the functional area.

Teaching achievements

At least 4 A1 drawings per person, including:
1. City and environment: General plan 1 : 500, overall aerial view, axonometric drawing.
2. Basic spatial expression: Plan, elevation and section 1:200–1:400.
3. Spatial analysis and expression: Conceptual analysis diagram, spatial composition analysis diagram, axonometric analysis diagram, sectional perspective view (no less than 2, which must include the sectional perspective view of large space and public space), indoor and outdoor human eye perspective views.
4. Manual model: Each tutor group makes a 1:500 general layout volume model, and each student makes a 1:500 concept block model.

Teaching progress

This design course contains 8 weeks in total.
Week 1: Teaching (1 class hour), research sites and cases, make site models (SU model + physical model 1:500) and collect relevant case data.
Week 2: Students collect case reports and discuss preliminary conceptual plans (including the layout of the relationship between the block and the site, and the basic layout of the interior space).
Week 3: Concept deepening, perfect the preliminary architectural functional layout and spatial form scheme (including basic spatial units and their combinations), and making a spatial form model.
Week 4: The plan is finalized, and the space skin, plan function, and block environment pattern are defined.
Week 5: Plan deepening I, the space skin, plan function, and block environment are deepened.
Week 6: Plan deepening II, refining of skin, design of theater space and other important public spaces.
Week 7: Schematic expression, complete the plan, elevation and section drawings, and improve the SU design model.
Week 8: Draw, typeset.

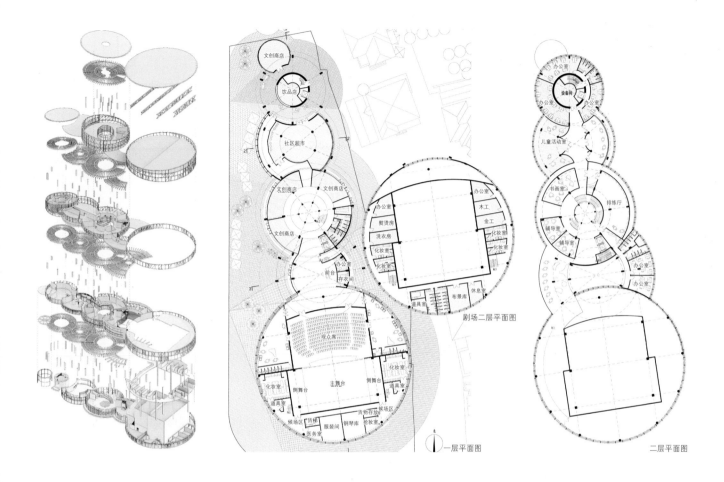

学生：高禾雨，王梓蔚 Students: GAO Heyu, WANG Ziwei

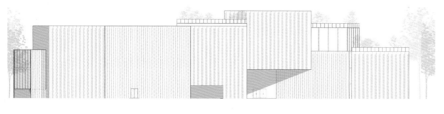
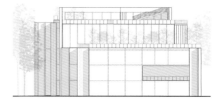
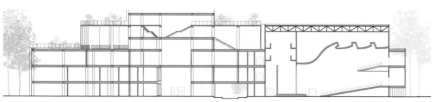

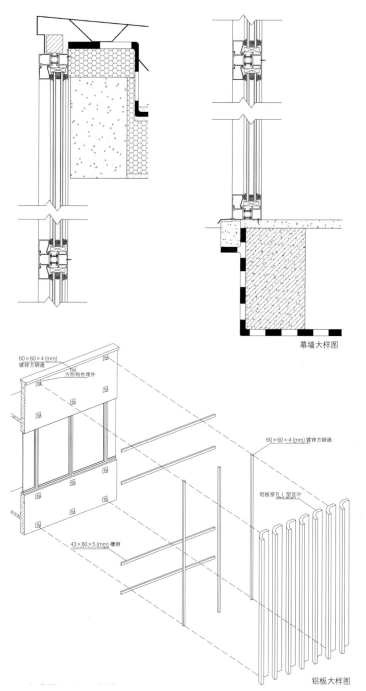

幕墙大样图

铝板大样图

学生：夏月，顾靓　Students: XIA Yue, GU Liang

实践课程 PRACTICAL COURSES

工 地 实 习
PRACTICE IN CONSTRUCTION PLANT

傅筱 吴佳维
FU Xiao, WU Jiawei

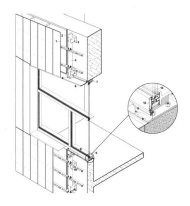

教学目标

工地实习的训练目的是加深学生对建筑从图纸到实际建造过程的认识和理解，重点理解图纸与建造的关联。通过课程了解建筑外围护结构常用构件的材料与构造，了解相关的建材生产和施工基本流程，掌握建筑技术图纸的表达规范和绘制方法，为后续的高年级专业学习以及研究生阶段的研究打下一定的技术知识基础。

人员组成

学生：2019级19人，实际参加人数19人。
任课教师：傅筱教授，吴佳维副研究员。
现场指导：建材厂技术人员。
助教：研究生卢卓成。

实践时间及地点

实习时间：2022年6月至8月，建材厂参观2天，研究及绘图约10天。
参观地点：南京六合丰彩门窗厂、南京高淳奥捷GRC厂、宜兴市和特陶砖厂。

教学内容

课堂讲授：由教师课堂讲授相关的技术知识。
厂家调研：教师和助教带领学生在建材厂进行考察和学习。
知识梳理：结合调研收集到的资料和教师的讲解，梳理从原材料到建材生产、现场安装的全过程。
构造轴测图绘制：进一步搜索图集、案例，绘制相关节点的构造轴测图。
实习报告：根据厂家考察和课堂讲授的学习后，完成实习报告。

教学成果

本次以建材调研为形式的工地实习是继"建筑技术一：建构设计"课程与"古建筑测绘实习"课程后对建造议题的再一次专题训练。建材厂实地调研生动地向学生们展示了常用建筑外围护材料从原料到产品的工艺原理和生产过程，使他们吸收到许多书本以外的知识。学生在了解材料特性后，进一步搜集案例、查阅资料，在教师的引导下完成了"原材料—生产过程—产品—现场安装"的信息梳理，同时绘制典型案例中的关键节点构造轴测图。经过教师几轮点评和改正，大部分学生掌握了GRC板材、窗、陶板、陶砖和陶棍等立面材料构件与建筑土建部分的连接方式，能够根据节点的平、剖面图纸绘制轴测图，正确表达各个构件的空间关系。此外，本次课程在节点研究中鼓励学生使用BIM软件以建立构件装配式思维；在输出成果的格式上特别注意版面、字体和线型等的规范和统一，综合训练了图纸表达能力。课程结束后，以展览的形式向全院师生汇报了教学的过程和成果。

Teaching objectives

The training purpose of the construction site practice is to deepen students' knowledge and understanding of the construction process from drawings to actual construction, focusing on understanding the relationship between drawings and construction. Through the course, students can understand the materials and structures of the commonly used components of the building envelope, understand the basic process of building material production and construction, master the expression norms and drawing methods of architectural technical drawings, and lay a solid technical knowledge base for the subsequent professional study in senior years and postgraduate research.

Staff composition
Students: 19 people (2019), the actual number of participants is 19 people.
Teachers: Professor Fu Xiao, Associate Researcher Wu Jiawei.
On-site guidance: Technical staffs of building materials factory.
Teaching assistant: Graduate student Lu Zhuocheng.

Practice time and places
Internship time: From June to August, 2022, visit the building materials factory for 2 days, research and draw for about 10 days.
Visiting places: Nanjing Liuhe Fengcai Door and Window Factory, Nanjing Gaochun Aojie GRC Factory, Yixing Hete Pottery Brick Factory.

Teaching contents
Lectures: Teachers teach relevant technical knowledge in class.
Factory research: Teachers and teaching assistants lead students to have study tours in the building materials factory.
Knowledge organization: Organize the whole process from raw materials to building material production and on-site installation combined with the information collected in the research and teacher's explanation.
Structural axonometric drawing: Further search for atlases and cases, and draw structural axonometric drawings of relevant nodes.
Internship report: According to the study tours and lectures, complete the internship report.

Teaching achievements
This construction site practice in the form of building materials research is another special training on construction issues following the courses of "Architectural Technology 1: Structural Design" and "Practice of Ancient Architecture Surveying and Mapping". The on-the-spot survey of the building materials factory vividly demonstrated to the students the technological principles and production process of commonly used building envelope materials from raw materials to products, enabling them to absorb a lot of knowledge beyond books. After understanding the characteristics of materials, students further collected cases and searched for information. Under the guidance of teachers, they completed the information organization of "raw materials—production process—products—on-site installation", and at the same time drew the structural axonometric diagrams of key nodes in typical cases. After several rounds of comment and correction by teacher, most of the students have mastered the connection methods of facade material components such as GRC plates, windows, terracotta panels, terracotta bricks and terracotta sticks with the civil engineering parts of the building, and are able to draw axonometric according to the plan and section drawings of the nodes. Figure, correctly express the spatial relationship of each component. In addition, this course encourages students to use BIM software to establish component assembly thinking in node research; pay special attention to the standardization and unification of the layout, font and line type, etc. in the format of output results, which comprehensively trains the ability to express drawings. After the course, the Teaching progress and results were reported to the teachers and students of the school in the form of an exhibition.

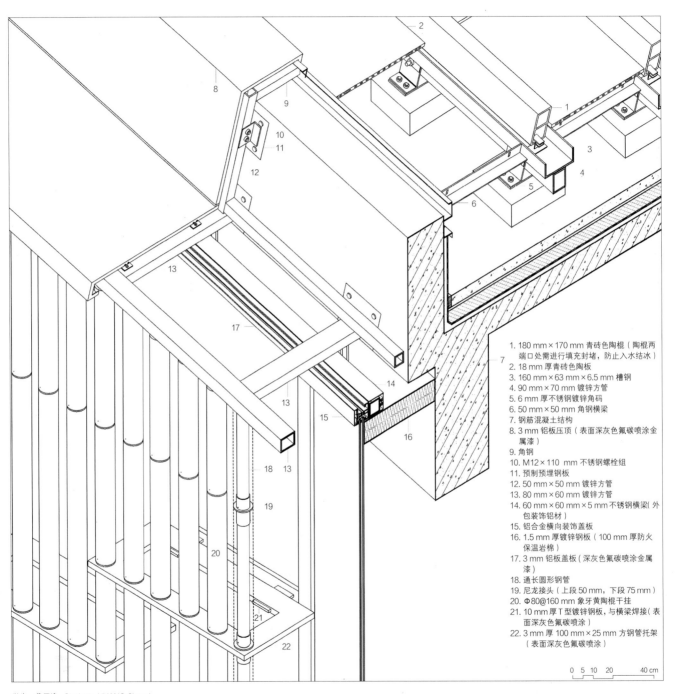

1. 180 mm×170 mm 青砖色陶棍（陶棍两端口处需进行填充封堵，防止入水结冰）
2. 18 mm 厚青砖色陶板
3. 160 mm×63 mm×6.5 mm 槽钢
4. 90 mm×70 mm 镀锌方管
5. 6 mm 厚不锈钢镀锌角码
6. 50 mm×50 mm 角钢横梁
7. 钢筋混凝土结构
8. 3 mm 铝板压顶（表面深灰色氟碳喷涂金属漆）
9. 角钢
10. M12×110 mm 不锈钢螺栓组
11. 预制预埋钢板
12. 50 mm×50 mm 镀锌方管
13. 80 mm×60 mm 镀锌方管
14. 60 mm×60 mm×5 mm 不锈钢横梁（外包装饰铝材）
15. 铝合金横向装饰盖板
16. 1.5 mm 厚镀锌钢板（100 mm 厚防火保温岩棉）
17. 3 mm 铝板盖板（深灰色氟碳喷涂金属漆）
18. 通长圆形钢管
19. 尼龙接头（上段 50 mm，下段 75 mm）
20. Φ80@160 mm 象牙黄陶棍干挂
21. 10 mm 厚 T 型镀锌钢板，与横梁焊接（表面深灰色氟碳喷涂）
22. 3 mm 厚 100 mm×25 mm 方钢管托架（表面深灰色氟碳喷涂）

学生：黄辰逸 Student: HUANG Chenyi

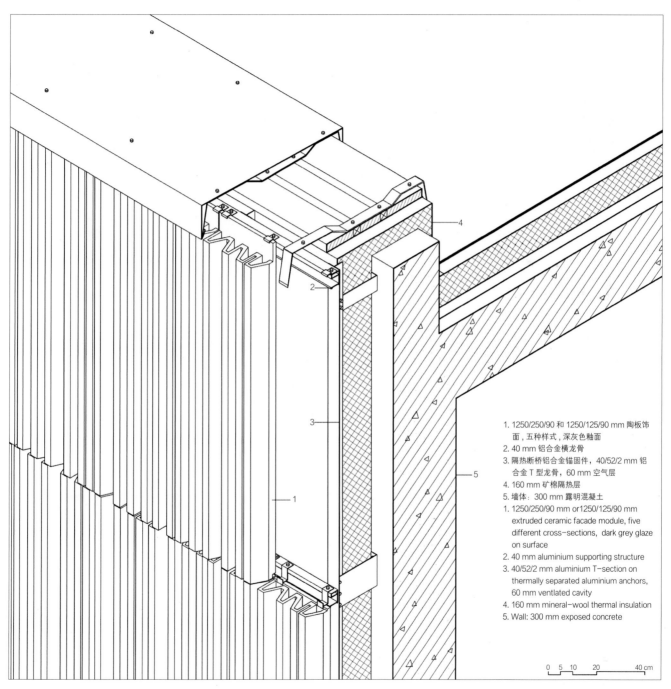

1. 1250/250/90 和 1250/125/90 mm 陶板饰面，五种样式，深灰色釉面
2. 40 mm 铝合金横龙骨
3. 隔热断桥铝合金锚固件，40/52/2 mm 铝合金T型龙骨，60 mm 空气层
4. 160 mm 矿棉隔热层
5. 墙体: 300 mm 露明混凝土

1. 1250/250/90 mm or 1250/125/90 mm extruded ceramic facade module, five different cross-sections, dark grey glaze on surface
2. 40 mm aluminium supporting structure
3. 40/52/2 mm aluminium T-section on thermally separated aluminium anchors, 60 mm ventlated cavity
4. 160 mm mineral-wool thermal insulation
5. Wall: 300 mm exposed concrete

学生：石珂千 Student: SHI Keqian

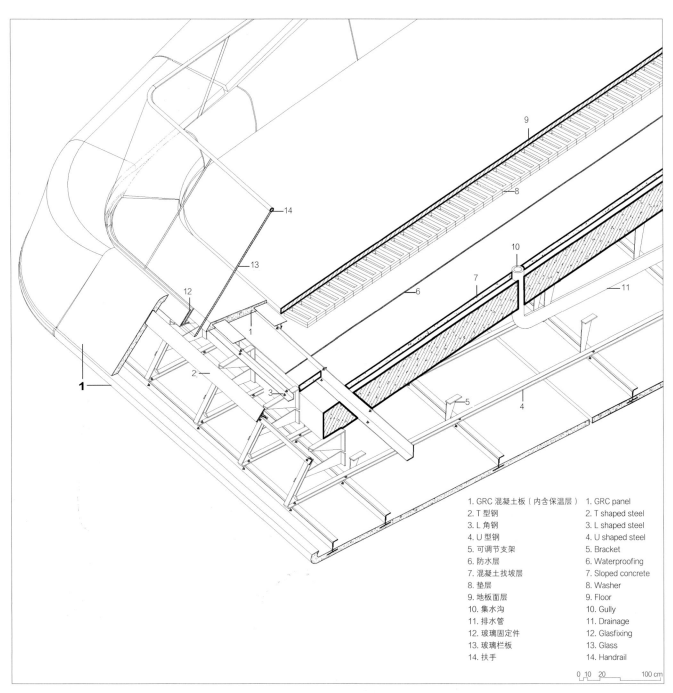

学生：高赵龙 Student: GAO Zhaolong

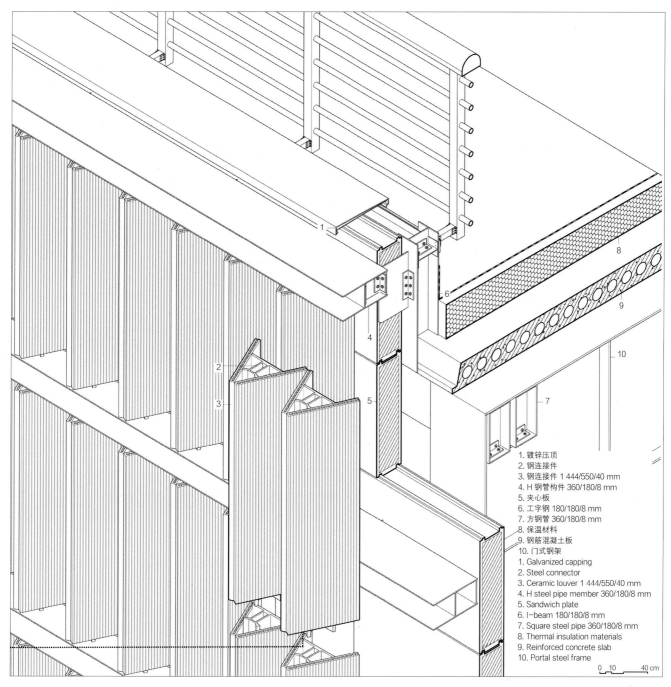

学生：邱雨婷 Student: QIU Yuting

建筑设计（七）ARCHITECTURAL DESIGN 7
高层办公楼设计
DESIGN OF HIGH-RISE OFFICE BUILDINGS

吉国华 王铠 尹航
JI Guohua, WANG Kai, YIN Hang

教学目标
建筑设计七共分为三组，大致提供三个主要的建筑发展方向：性能化导向建筑设计、景观导向的高层建筑、复合功能的立体城市。

本次课程设计首先希望同学掌握高层建筑的基本特点，研究当代高层建筑的设计策略，了解高层建筑涉及的相关规范与知识，提高综合分析并解决问题的能力。其次还希望同学能主动将建筑与周边文脉、景观等环境要素有机结合，研究环境，在建筑方案中预设某种设计策略，充分发挥创造性，尝试进行某种绿色摩天楼的设计。

教学内容
1. 场地与相关指标
本次课程设计的场地位于在南京市南部新城大校场地块，大校场机场跑道公园是大校场机场搬迁后对跑道整体保护形成的新型城市景观廊道。在新一轮规划中，跑道北侧为城市文化客厅，将连续布置大体量多层文化建筑，南侧则布置高层混合功能建筑。

本次课程设计提供轨道五号线北侧三组相邻地块供选择进行设计，三组地块都位于跑道景观廊道的东入口，自东向西控高分别为200~400 m和150~200 m不等。同学们可在对上位规划进行研读后，根据不同地块的特点进行设计。

各地块建筑容积率在3（高层）或4（超高层）以上。具体指标见资料图件、相关退让间距等指标可参考资料图件及《南京市城市规划条例实施细则》等。

2. 功能与交通
设计地块在上位规划中确定的性质有办公与商住两种，课程拟减少功能限制，由学生自主调研与策划地块的功能，可设置多种复合功能，具体面积比例自定；合理组织车行、人行入口流线，按不同功能设置出入口集散空间；合理组织场地交通与周边道路关系，合理设置停车场，停车数量（机动车与非机动车）应尽量满足规范要求，地下停车场出入口大于2个。

3. 相关规范
《民用建筑设计通则》（GB 50352—2005）
《城市道路和建筑物无障碍设计规范》（JGJ 50—2001）
《办公建筑设计规范》（JGJ 67—2006）
《车库建筑设计规范》（JGJ 100—1998）
《建筑设计防火规范》（GB 50016—2014）
《汽车库、修车库、停车场设计防火规范》（GB 50067—1997）
《南京市建筑物配建停车设施设置标准与准则》

时间安排（8周）
第一周：分析场地（1∶1000场地环境模型），研究设计策略。
第二周：推敲设计策略（1∶1000模型）。
第三周：制作概念模型（1∶1000模型）。
第四周：总平面设计（草图、1∶500模型）。
第五周：平面、立面与细部深化设计。
第六周：制作技术图纸、排版。
第七周：完成技术图纸，制作表现图、表现模型。
第八周：完成表现图、模型。

Teaching objectives
Architectural Design 7 is divided into three groups, providing roughly three main architectural development directions: performance-oriented architectural design, landscape-oriented high-rise buildings, and composite functional three-dimensional cities.

This course design first hopes that students can master the basic characteristics of high-rise buildings, study the design strategies of contemporary high-rise buildings, understand the relevant norms and knowledge involved in high-rise buildings, and improve the ability of comprehensive analysis and problem solving.Second, it also hopes that students can take the initiative to organically integrate the building with surrounding cultural heritages, landscapes and other

  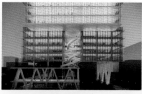 

environmental elements, study the environment, preset a certain design strategy in the architectural scheme, give full play to creativity, and try to carry out some kind of green skyscraper design.

Teaching contents

1. The site and related indicators

The site designed for this course is located in the Dajiaochang in the south of Nanjing City. Dajiaochang Airport Runway Park is a new urban landscape corridor formed with the overall protection of the runway after the relocation of Dajiaochang Airport. In the new round of planning, the north side of the runway will be the urban cultural living room, which will be continuously arranged with lots of large multi-storey cultural buildings, and the south side will be arranged with a high-rise mixed-function buildings.

This course design provides three groups of adjacent plots on the north side of Track Line 5 for selection and design. The three groups of plots are located at the east entrance of the runway landscape corridor, and the control heights from east to west are 200–400 m and 150–200 m respectively. Students can design according to the characteristics of different plots after studying the upper planning. The floor area ratio of each plot is above 3 (high-rise) or 4 (super high-rise). Specific indicators can be seen in the data map, and relevant indicators such as the concession distance can be referred to the data map and the Implementation Rules of the Nanjing Urban Planning Regulations.

2. Functions and transportation

The properties of the design plots determined in the upper planning are office buildings and commercial residences. The course plans to reduce the functional restrictions, and students will independently investigate and plan the functions of the plots. Multiple composite functions can be planned, and the specific area ratio is determined by themselves. Reasonably organize the vehicle and pedestrian entrance flow lines, according to different functions entrance and exit distribution space is set up. The relationship between the site traffic and the surrounding roads should be reasonably organized, and the parking lot should be reasonably set up. The number of parking spaces (motor vehicles and non-motor vehicles) should meet the requirements of local laws and regulations as far as possible, and the underground parking lot should have more than 2 entrances and exits.

3. Relevant specifications

Code for Design of Civil Buildings (GB 50352—2005)
Code for Design on Accessibility of Urban Roads and Buildings (JGJ 50—2001)
Code for Design of Office Buildings (JGJ 67—2006)
Code for Design of Parking Garage Buildings (JGJ 100—1998)
Code for Fire Protection Design of Buildings (GB 50016—2014)
Code for Fire Protection Design of Garage, Motor Repair shop and Parking Area (GB 50067—1997)
Standards and Guidelines for the Installation of Parking Facilities of Buildings in Nanjing

Schedule (8 weeks)

Week 1: Analysis the site (1∶1 000 site environment model), research the design strategy.
Week 2: Review the design strategy (1∶1 000 model).
Week 3: Make the conceptual model (1∶1 000 model).
Week 4: General graphic design (sketch, 1∶500 model)).
Week 5: Plan, elevation and detailed in-deep design.
Week 6: Make technical drawings and typeset.
Week 7: Complete the technical drawing, make the presentation drawing and the presentation model.
Week 8: Complete the presentation map and the model.

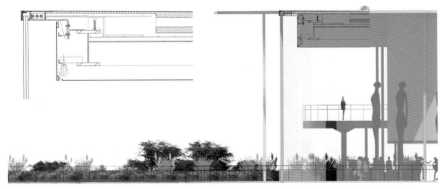

学生：张百慧，张新雨　Students: ZHANG Baihui, ZHANG Xinyu

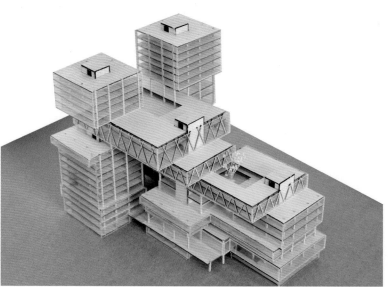

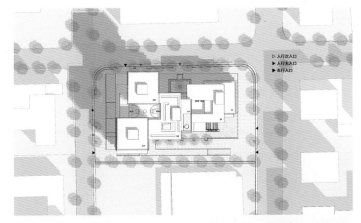

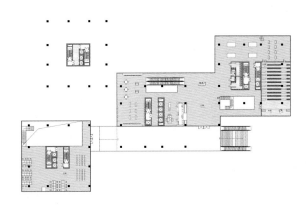

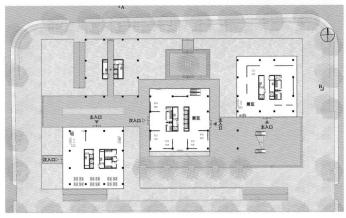

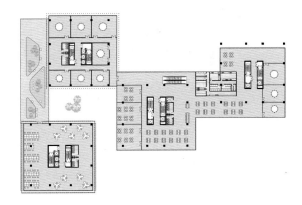

学生：陈锐娇，宋佳艺　Students: CHEN Ruijiao, SONG Jiayi

建筑设计（八）ARCHITECTURAL DESIGN 8

城市设计
URBAN DESIGN

童滋雨　胡友培　唐莲
TONG Ziyu, HU Youpei, TANG Lian

课程内容
计算化城市设计

教学目标
中国的城市发展已经逐渐从增量扩张转向存量更新。通过对城市建成环境的更新改造而提升环境性能和质量，将成为城市建设的新热点和新常态。与此同时，5G、物联网、无人驾驶等技术的发展又给城市环境的使用方式带来了新的变化。如何在城市更新设计中拓展建筑设计的边界也就成了新的挑战。

城市更新不但需要对建成环境本身有更充分的认知，也要对其中的人流、车流乃至水流、气流等各种动态的活动有正确的认知。从设计上来说，这也大大提高了设计者所面临的问题的复杂性，仅靠个人的直观感受和形式操作难以保证设计的合理性。而借助空间分析、数据统计、算法设计等数字技术，我们可以更好地认知城市形态的特征，理解城市运行的规则，并预测城市未来的发展。通过规则和算法来计算生成城市也是对城市设计思维范式的重要突破。

因此，本次设计将针对这些发展趋势，以城市街巷空间为研究对象，通过思考和推演探索其更新改造的可能性。通过本次设计，学生们可以理解城市设计的相关理论和方法，掌握分析城市形态和创造更好城市环境质量的方法。

设计场地
课程设计范围位于南京市鼓楼区广州路两侧，东至中央路，西至上海路，总长度约750 m，南北边界可根据调研自己确定。

成果要求
本次设计以小组为单位，每小组2人。每组成果包括8张A1图纸和1份A4成果文本，具体内容可包括但不限于以下部分。
1. 设计表达：平面图、立面图、轴测图、透视图等。
2. 设计推演：设计形成过程的分析图。
3. 设计评估：对设计成果的各种评估分析。

教学进度
阶段一：场地调研与案例分析（2周）。
阶段二：设计目标确定与总体布局方案（2周）。
阶段三：方案完善与局部深化（3周）。
阶段四：制图与排版（1周）。

Course content
Computerized urban design

Teaching objectives
China's urban development has gradually shifted from incremental expansion to stock renewal. Improving environmental performance and quality through the renewal and transformation of urban built environment will become a new hot

spot and new normal of urban construction. At the same time, the development of 5G, Internet of things, unmanned driving and other technologies has brought new changes to the use of urban environment. How to expand the boundary of architectural design in urban renewal design has become a new challenge.

Urban renewal not only needs to have a better understanding of the built environment itself, but also needs a correct understanding of the pedestrian stream, vehicle stream, water stream, air stream and other dynamic activities. In terms of design, it also greatly improves the complexity of problems faced by designers. It is difficult to ensure the rationality of design only by personal intuitive feelings and formal operations. With the help of various digital technologies such as spatial analysis, data statistics and algorithm design, we can better understand the characteristics of the urban form, understand the rules of the urban operation, and predict the future development of the city. Calculating and generating cities through rules and algorithms is also an important breakthrough in the thinking paradigm of urban design.

Therefore, this design will aim at these development trends, take the urban street space as the research object, and explore the possibility of its renewal and transformation through thinking and deduction. Through this design, students can understand the relevant theories and methods of urban design, and master the methods of analyzing the urban form and creating better urban environmental quality.

Design site
The course design site is located on both sides of Guangzhou Road, Gulou District, Nanjing, east to Zhongyang Road and west to Shanghai Road, with a total length of about 750 m. The north-south boundary can be determined according to the investigation.

Achievement requirements
This design takes the group as the unit, with 2 people in each group. Each group's achievements include 8 A1 drawings and 1 A4 achievement text. The specific contents can include but are not limited to the following parts:
1. Design expression: Plan, elevation, axonometric drawings, perspective view, etc..
2. Design deduction: Analysis diagram of design formation process.
3. Design evaluation: Various evaluation and analysis of design results.

Teaching progress
Stage 1: Site investigation and case analysis (2 weeks).
Stage 2: Determination of design objectives and the overall layout scheme (2 weeks).
Stage 3: Scheme improvement and local deepening (3 weeks).
Stage 4: Drawing and typesetting (1 week).

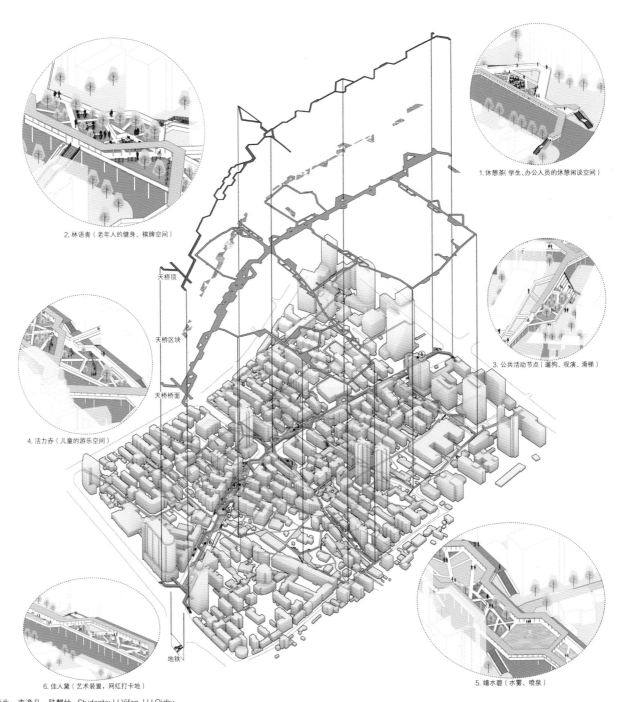

1. 休憩茶（学生、办公人员的休憩闲谈空间）
2. 林语青（老年人的健身、棋牌空间）
3. 公共活动节点（遛狗、观演、滑梯）
4. 活力赤（儿童的游乐空间）
5. 嬉水碧（水雾、喷泉）
6. 佳人黛（艺术装置，网红打卡地）

学生：李逸凡，陆麒竹 Students: LI Yifan, LU Qizhu

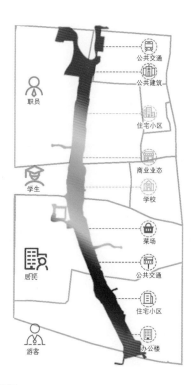

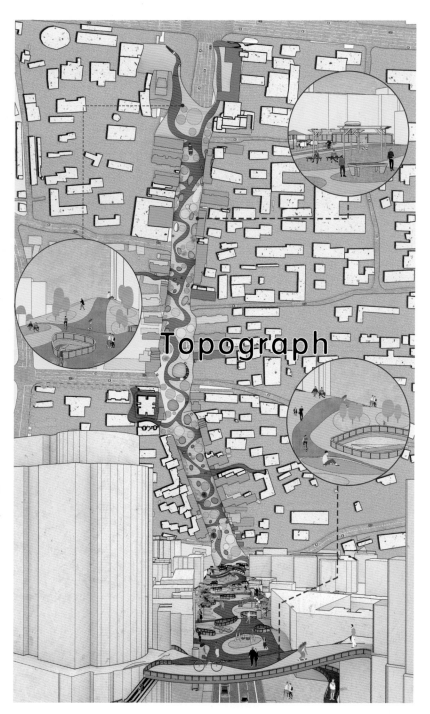

Topograph

业态等级划分

业态等级划分

学生：顾祥姝，田舒琳　Students: GU Xiangshu, TIAN Shulin

本科毕业设计 GRADUATION PROJECT

废墟的可能性：国民政府中央广播电台旧址附属建筑及环境改造
THE POSSIBILITY OF RUINS: ANNEX BUILDINGS AND ENVIRONMENTAL RENOVATION OF THE FORMER SITE OF THE NATIONAL GOVERNMENT CENTRAL RADIO AND STATION

华晓宁
HUA Xiaoning

课程介绍

建筑学意义上，废墟的可能性可以从双向维度来解析——废+墟。

废，是人对废墟的使用状态，包括：1.原点状态，观看而非使用废墟，等于一种归零状态；2.废墟被重新注入新的固定使用功能；3.废墟处于一种空的状态，包容多种临时使用功能。

墟，是废墟物的存在状态，包括：1.原点状态，即废墟的现状保留；2.修复，通过修复和加固，使废墟变得重新坚固和崭新；3.墟化，通过主动设计来强化废墟的废墟化状态，把废墟里面隐藏的深层次内涵剥离显现出来，使之变成更加纯粹的废墟。

上述双向维度交叉在一起，就形成了一个"废墟坐标系"，人的使用状态是一个轴，物的存在状态是一个轴，中间的原点是观看和保留。以此为基础，可以引申出"废墟意向设计"的方法论体系。

本次课程设计以国民政府中央广播电台旧址为改造设计对象，要求学生初步接触废墟的基本概念，探讨把废墟的思想理念纳入旧建筑改造的设计过程中，通过废墟意向设计和重点空间设计，拓展设计思维，为今后的旧建筑改造设计实践打下基础。进一步，可以延展思考有关"废墟建筑学"的理论。这样的深度思考将有助于学生反思、回顾和总结已有的建筑学知识，开拓视野，为未来进入更加广阔的学术领域，应对更多的可能性和不确定的未来做好准备。

Course descriptions

In the sense of architecture, the possibility of ruins can be analyzed from the two-way dimension—waste + remaining parts.

Waste is the use state of ruins, including: 1. The origin state, viewing rather than using the ruins, equal to a state of zero; 2. The ruins are re-injected with new fixed functions; 3. The ruins are in an void state and contain a variety of temporary use functions.

Remaining parts are the existing state of ruins, including: 1. The origin state, that is, the status quo of ruins; 2. Repair, to make the ruins strong and new through restoration and reinforcement; 3. Fall into ruins, through the active design to strengthen the state of ruins, the hidden deep connotations of the ruins are stripped out and revealed, so as to make ruins more pure.

The above bidirectional dimensions are crossed to form a "ruin coordinate system", in which the use state of humans is an axis, the state of existence of things is an axis, and the origin in the middle is to watch and retain. On this basis, the methodology system of "intentional design of ruins" can be extended.

This course design takes the former site of the Central Radio and Station of the National Government as the object of renovation design. Students are required to have preliminary contact with the basic concept of ruins, explore the idea of incorporating ruins into the design process of old building renovation, and expand design thinking through the intention design of ruins and key space design, laying the foundation for the future old building renovation design practice. Further, the theory of "ruins architecture" can be extended. Such in-depth thinking will help students to reflect, review and summarize the existing knowledge of architecture, broaden their horizons, and prepare for entering a broader academic field and dealing with more possibilities and uncertainties in the future.

学生：孙穆群　Student: SUN Muqun

疗愈环境与"大健康"校园创新设计
HEALING ENVIRONMENT AND INNOVATIVE DESIGN OF "BIG HEALTH" CAMPUS

本科毕业设计 GRADUATION PROJECT

窦平平
DOU Pingping

题目背景

大学生的心理健康状况受到社会各方面的高度关注,所面临的身心健康问题也迫切需要校园提供适宜的疗愈场所。2020年9月,国家卫生健康委员会宣布各个高中及高等院校将抑郁症筛查纳入学生健康体检内容,建立学生心理健康档案,并重点关注测评结果异常的学生。多位专家还发现,目前抑郁症低龄化趋势正进一步显现。以抑郁症为代表的精神疾病正严重影响青少年身心健康,且有极端化倾向,应引起足够重视。

教学目标

本次毕业设计以真实项目"南京大学心理健康与研究中心改造设计"为依托,与南京大学高研院、社会学院、艺术学院等跨学科协作,以塑造面向"大健康"的校园环境为目标,提出一套基于南京大学的校园疗愈环境设计模式,并选点完成带有前瞻性和综合性的优化设计。

重点问题

1. 环境心理学:为疗愈环境的研究提供了理论基础,其研究重点与疗愈环境相契合,同样关注于环境对使用者的影响以及使用者的身心需求。
2. 循证设计:环境心理学和疗愈环境研究中的重要研究方法。
3. 景观疗愈与亲自然设计:积极调动身体五感,将环境设施与景观要素结合进行空间设计。
4. 声音疗愈与视听交互设计:环境中各个要素的存在,都对环境中个体的恢复产生积极的促进或消极的阻碍作用。声景作为重要的环境要素,具有社会和审美的属性,能够对恢复性感受产生直接的影响。

Title background

The mental health of university students is of great concern to all sectors of society, and the physical and mental health problems they face urgently require suitable places that campuses provide for healing. In September 2020, the National Health Commission announced that all high schools and colleges and universities will include depression screening in students' health examinations, establish students' mental health files, and focus on students with abnormal assessment results. Many experts also found that the current trend of depression at a younger age is further emerging. The mental illness represented by depression is seriously affecting the physical and mental health of teenagers, and has the tendency of extreme, which should be paid enough attention to.

Teaching objectives

This graduation project is based on the real project "Transformation Design of Mental Health and Research Center of Nanjing University", in collaboration with Institute for Advanced Studies in Humanities and Social Sciences, School of Sociology, School of Art, etc., in order to shape the campus environment for "big health" as the goal. A set of healing environment design model is proposed based on the Nanjing University campus, and points are selected to complete the forward-looking and comprehensive optimization design.

Key problems

1. Environmental psychology: Provide a theoretical basis for the study of the healing environment. The focus of the research is in line with the healing environment, which also focuses on the impact of the environment on users and users' physical and psychological needs.
2. Evidence-based design: Important research methods in environmental psychology and healing environment research.
3. Landscape healing and design close to nature: Actively mobilise the body's five senses, combine environmental amenities with landscape elements in spatial design.
4. Sound healing and audio-visual interaction design: The existence of various elements in the environment can positively promote or negatively hinder the recovery of individuals. As an important environmental element, soundscape has social and aesthetic properties and can have a direct impact on restorative feelings.

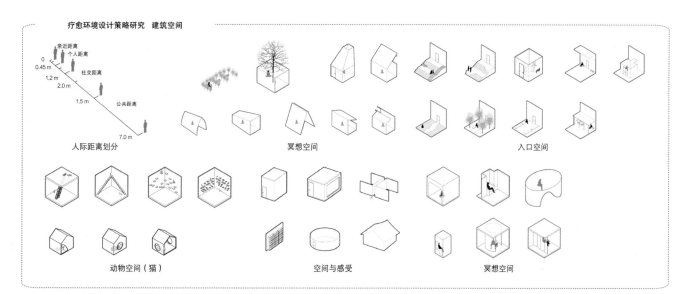

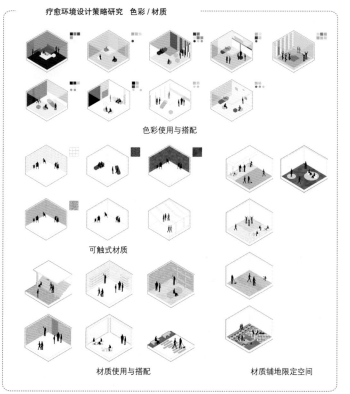

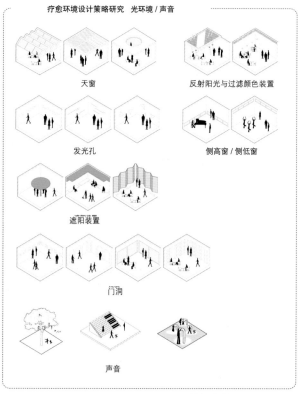

学生：沈洁　Student: SHEN Jie

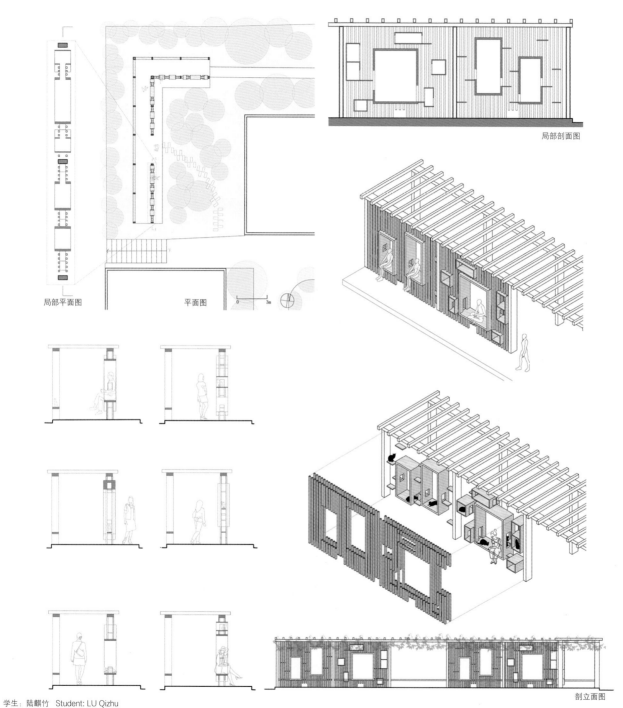

局部剖面图
局部平面图　平面图
剖立面图

学生：陆麒竹　Student: LU Qizhu

学生：沈洁 Student: SHEN Jie

亚太地区"干栏式"建筑研究:以日本列岛为例
STUDY ON THE "STILT STYLE" BUILDINGS IN ASIAN-PACIFIC REGION: TAKING THE JAPANESE ARCHIPELAGO AS AN EXAMPLE

赵潇欣
ZHAO Xiaoxin

研究背景
《中国建筑史》教材将中国古代建筑分类为:抬梁、穿斗与干栏式。然而,这里提到的"干栏式"是建筑与地面接触关系的空间形式,与前述抬梁、穿斗以结构形式的分类标准不在同一个层面。世界范围内也有众多国家的传统建筑采用"干栏式"的形式,如日本的舞台造(也称悬崖造)、澳洲的昆士兰风格建筑等。本次研究把中国建筑置于更大的亚太地区乡土建筑视野中,比较亚太地区其他国家"干栏式"乡土建筑的特点,以思考各地区干栏式建筑的空间特征、结构逻辑、建造特征、社会文化特征的联系与异同,更好地理解"干栏式"建筑作为一种空间形式,何以在世界范围内广泛应用,理解建筑文化的跨地域性。

研究问题
干栏式建筑在亚太地区的分布情况是怎样的?在具体某地区的分布情况是怎样的?某地区的干栏式乡土建筑的空间是怎样的?建造特点是怎样的?结合当地的气候、社会文化特征讨论,为什么当地会采用干栏式建筑?

Research background
The textbook *History of Chinese Architecture* classifies Chinese ancient buildings into: post and lintel construction, column and tie construction, stilt style. However, the "stilt style" mentioned here is the spatial form of the contact relationship between the building and the ground, which is not at the same level as the classification standard of the above-mentioned structural form of post and lintel construction, column and tie construction. There are also many countries around the world that use the "stilt style" form for traditional buildings, such as Japan's stage buildings (also known as cliff buildings), Australia's Queensland style architecture and so on. This study puts Chinese architecture in the larger perspective of vernacular buildings in the Asia-Pacific region, compares the characteristics of "stilt style" vernacular buildings in other countries in the Asia-Pacific region, and considers the spatial characteristics, structural logic, construction characteristics, and social and cultural characteristics of stilt style buildings in different regions, so as to better understand stilt style buildings as a spatial form why stilt style buildings are widely used in the world, and understand the trans-regional nature of architectural culture.

Research problems
What is the distribution of stilt style buildings in the Asia Pacific region? What is the distribution in a particular region? What is the space of local buildings in a certain area? What are the construction characteristics? Considering the local climate and social and cultural characteristics, why is the stilt style buildings adopted by locals?

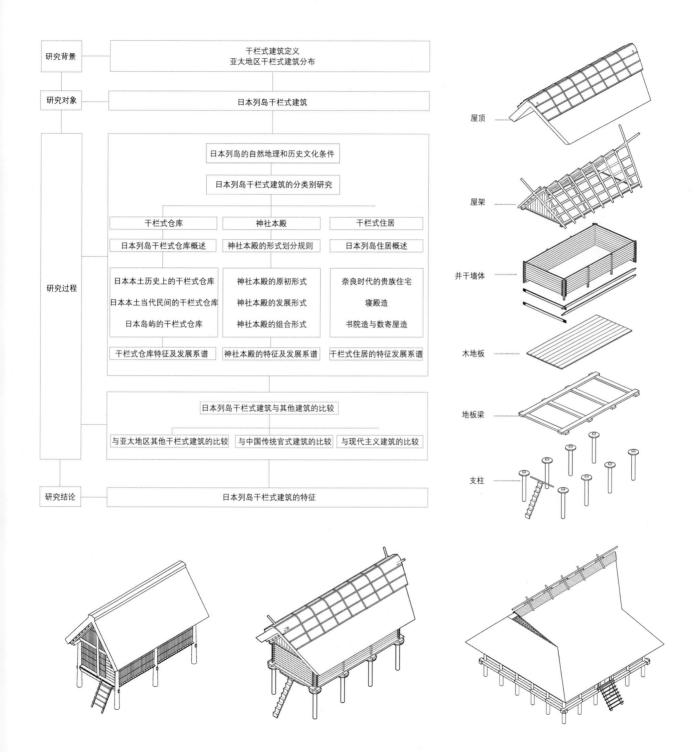

名称	大社造	住吉造
代表案例	出云大社本殿（现本殿）	住吉大社本殿

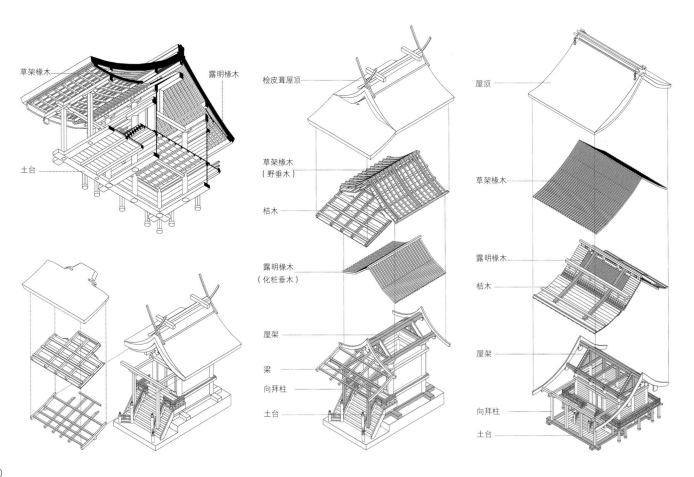

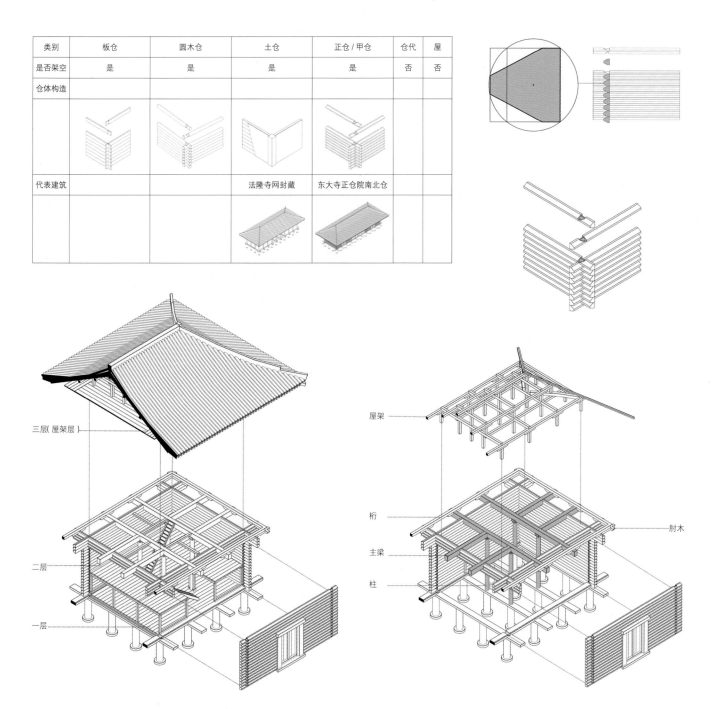

学生：罗宇豪　Student: LUO Yuhao

建筑设计研究（一）ARCHITECTURAL DESIGN RESEARCH 1

基本设计
BASIC DESIGN

周 凌
ZHOU Ling

研究课题
　　西连岛北片区更新改造规划及建筑设计

课程简介
　　连岛位于江苏、山东两省沿海交界处海域，行政建制隶属连云港市连云区，与连云港城区隔海相望。连岛东西长约 5.5 km，南北最宽处约 1.8 km。海岸线长约 18 km，礁石岸线约 8 km。连岛地形陡峭，起伏较大，山形错落有致。岛上植物种类丰富，植被密布，林地面积超过 4.5 万 km²，森林覆盖率达 65%。
　　本次设计分为片区更新改造规划与单体建筑设计两个部分，旨在培养学生依托生态、高效的片区更新理念，对单体建筑进行深入设计的能力。同时学习生态规划、业态策划、产业布局等知识。

设计内容
　　3~4 人一组进行设计：
　　任务一：片区更新改造规划；
　　任务二：单体建筑设计。
　　（四选一：城市客厅、海滨精品酒店集群、半岛星级酒店、山上民居组团）

学习内容
　　片区更新、建筑策划、总图布局、生态分析、用地分析、交通分析；分析图制作、效果图制作、动画制作等。

设计成果
　　每组 8 张 A1，包含总平面图、效果图、分析图等，设计说明 1 000 字，PPT 陈述，动画 2~5 min。

教学进度
　　第 1 周：现状分析与案例研究；
　　第 2 周：策划方案与概念规划；
　　第 3 周：概念设计；
　　第 4 周：概念设计；
　　第 5~6 周：建筑设计；
　　第 7~8 周：成果表达制作。

Research topic
Renovation planning and architectural design of the northern area of West Liandao

Course descriptions
Liandao is located in the sea area at the junction of the coastal areas of Jiangsu and Shandong provinces. Its administrative system is subordinate to Lianyun District of Lianyungang City, and it faces Lianyungang City across the sea. Liandao is about 5.5 km long from east to west, and about 1.8 km wide at its widest place from north to south. The coastline is about 18 km long, and the reef shoreline is about 8 km. The terrain of Liandao is steep, with large ups and downs, and the mountain forms are well arranged and proportioned. The island is rich in plant species and densely covered with vegetation. The forest area exceeds 45 000 km^2, and the forest coverage rate reaches 65%.
This design is divided into two parts: area renewal planning and individual building design, aiming at cultivating students' ability to carry out in-depth design of individual buildings relying on the concept of ecological and efficient area renewal. At the same time, they can learn ecological planning, business planning, industrial layout and other knowledge.

Design contents
Design in groups of 3–4:
Task 1: Area renewal planning;
Task 2: Individual building design.
(Choose one from four: City living room, seaside boutique hotel cluster, peninsula star hotel, mountain dwelling group)

Learning contents
Area renewal, architectural planning, overall layout, ecological analysis, land use analysis, and traffic analysis; Analysis drawing production, rendering production, animation production, etc.

Design achievements
Each group consists of 8 A1 sheets, including a general plan, rendering, analysis, etc., with a design description of 1 000 words, a PPT presentation, and an animation lasting 2–5 minutes.

Teaching progress
Week 1: Current situation analysis and case studies;
Week 2: Planning and conceptual planning;
Week 3: Conceptual design;
Week 4: Conceptual design;
Week 5–6: Architectural design;
Week 7–8: Production of results expression.

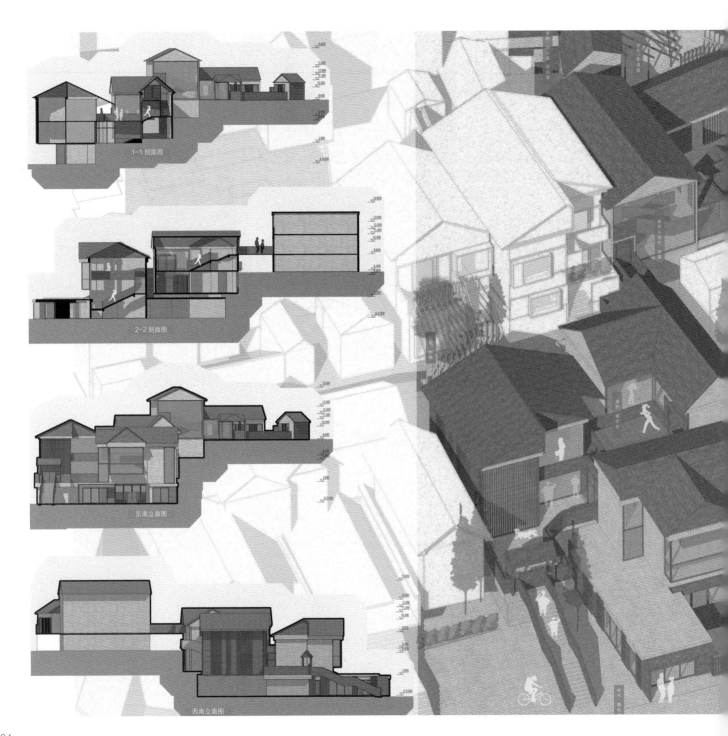

1-1 剖面图

2-2 剖面图

东南立面图

西南立面图

学生：王晓茜 Student: WANG Xiaoxi

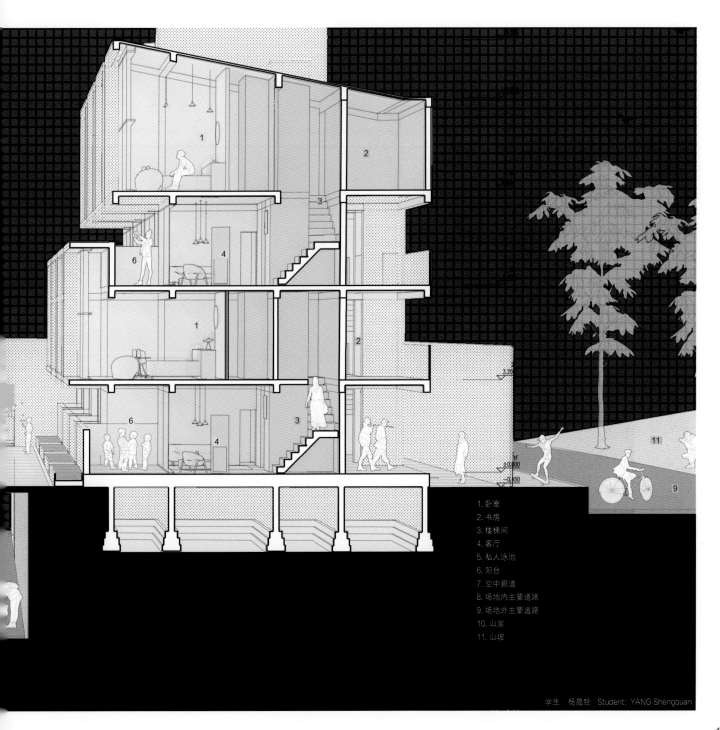

建筑设计研究（一）ARCHITECTURAL DESIGN RESEARCH 1

基本设计
BASIC DESIGN

傅 筱
FU Xiao

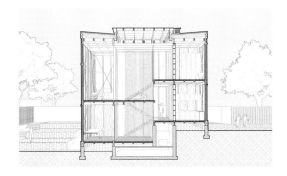

建筑要求

1. 建筑基底面积不得超过 130 m²。要求内部布局紧凑经济，使用功能合理，在满足功能需求之下，尽量减少面积以节省造价，总建筑面积不宜超过 220 m²。

2. 可选用结构为：钢筋混凝土框架结构（或剪力墙结构）、砖混结构、轻钢龙骨结构体系（或轻钢体系）、轻木框架结构体系（或轻木龙骨体系），同一地块的小组不得选用相同的结构体系。外墙材料选择需与结构体系有一定的关联性，并需要考虑保温隔热要求。

3. 主要入户空间要求朝南或者朝东。需配备服务入口，服务入口朝向自定。

4. 空调形式为小型一拖多 VRV 空调，需设计放置位置，并考虑室内出风口位置。

5. 明厨明卫，需进行烟囱设计。

6. 具体房间数量要求、面积自定：
（1）客厅；（2）餐厅；（3）厨房；（4）主卧 1 间（带卫生间，考虑小孩同居，步入式衣橱）；（5）客卧 1 间（使用公用卫生间）；（6）客卧 1 间（使用公用卫生间，未来作为子卧）；（7）丈夫工作空间；（8）公用卫生间，根据需要确定数量；（9）洗衣空间；（10）储藏空间。

技术要求：鼓励用 BIM（Revit）、Enscape 设计和表达。

参考书目：《加拿大木框架房屋建筑》（学院资料室），各类住宅设计书籍。

教学进度

1. 场地分析

时间：第一周。

内容：分析场地和相关案例研究，并提交分析报告和 1：100 场地模型（全组 1 个，用于平时方案讨论，场地范围由学生自行研究确定）。每组分析 2~3 个案例（需包含所选定的结构体系），组与组之间的案例尽量不要重复。

2. 了解基本建筑设计原理

时间：第二周。

内容：汇报分析报告，并提出初步概念方案，以工作模型或者图纸进行研究，需制作 PPT 汇报。

3. 组织空间与行为

时间：第三至六周。

内容：形成解决方案，以工作模型和图纸进行研究，需制作 PPT 汇报。

4. 设计研究与表达

时间：第七至八周。

内容：绘图。

成果要求

1. 完成平面图、立面图、剖面图图纸，平面图比例 1：50，剖面图、立面图比例 1：100，剖面图要求布置家具以及表达人的行为。

2. 表达空间关系的三维剖透视图 1~2 个，比例不低于 1：50，要求材料填充，必须有家具和人的行为表达。

3. 不小于 1：20 的表达设计意图的外墙大样，不少于 2 个。

4. 有利于表达形体和空间的透视图，可以是渲染图也可以是模型照片。

5. 其他有利于设计意图表达的图纸。

Building requirements

1. The building base area shall not exceed 130 m². It is required that the internal layout is compact and economical, and the use function is reasonable. Under the condition of meeting the functional requirements, the area should be reduced as much as possible to save the cost. The total construction area should not exceed 220 m².

2. The optional structures are: reinforced concrete frame structure (or shear wall structure), brick-concrete structure, light steel keel structure system (or light steel system), light wood frame structure system (or light wood keel system). Groups of the same block must not use the same structural system. The selection of exterior wall materials needs to have a certain relationship with the structural

system, and the thermal insulation requirements need to be considered.
3. The main entrance space is required to face south or east. It needs to be equipped with a service entrance, and the orientation of the service entrance can be determined by students.
4. The air conditioner is a small multi split VRV air conditioner, and the placement location needs to be designed, taking into account the location of the indoor air outlet.
5. The kitchen and bathroom, chimney design is required.
6. The specific number of rooms should have windows and the area are self-determined:
(1) Living room; (2) Dining room; (3) Kitchen; (4) 1 master bedroom (with bathroom, considering children will live together, walk-in closet); (5) 1 guest bedroom (use public bathroom); (6) 1 guest bedroom (public bathroom will be used, as a child bedroom in the future); (7) Husband's work space; (8) Public bathroom, the number is determined according to needs; (9) Laundry space; (10) Storage space.
Technical requirements: BIM (Revit), Enscape design and expression are encouraged.
Bibliography: *Canadian Timber Frame Housing Architecture* (College Reference Room), various residential design books.

Teaching progress
1. Site analysis
Time: Week 1.
Contents: Analyze the site and the related case studies, and submit the analysis report and 1:100 site model (a model for each group for scheme discussion, and the site scope should be determined by the students). Each group should analyze 2–3 cases (including the selected structural system), and the cases between groups should not be repeated.
2. Understand the basic architectural design principles
Time: Week 2.
Contents: Submit the analysis report, put forward a preliminary conceptual scheme, carry out the research based on the working model or drawings, and prepare a PPT report.
3. Organization space and behaviors
Time: Week 3–6.
Contents: Form a solution, carry out the research based on the working model or drawings, and prepare a PPT report.
4. Design research and expression
Time: Week 7–8.
Contents: Layouts.

Achievement requirements
1. Complete the plan, elevation and section drawings, the scale of the plan is 1 : 50, the scales of the section, and the elevation are 1 : 100. The section requires the arrangement of furniture and the expression of human behaviors.
2. 1–2 three-dimensional cross-sectional perspective views expressing spatial relationships, with a scale of no less than 1 : 50, requiring materials to be filled, and furniture and human behaviors must be expressed.
3. No less than 2 large samples of exterior walls expressing design intentions in a scale of no less than 1 : 20.
4. A perspective view that is conducive to expressing the shape and space, which can be a rendering or a model photo.
5. Other drawings that are conducive to the expression of design intentions.

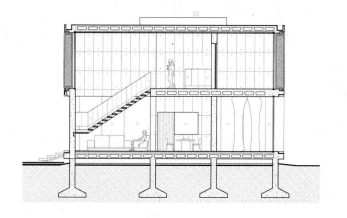
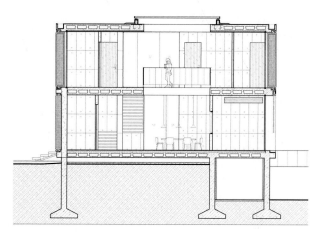
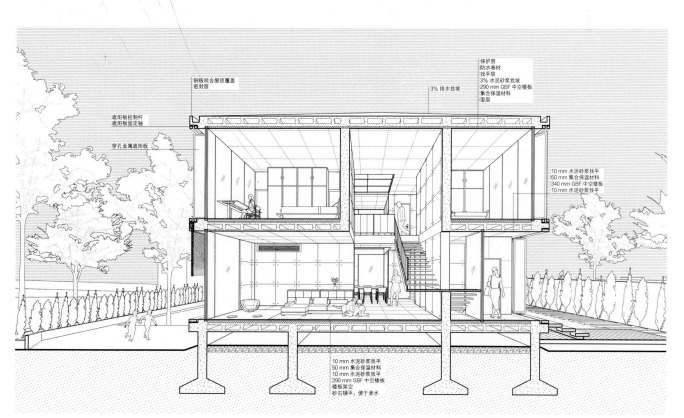

学生：孙珂，张鹏，崔沂瑄　Students: SUN Ke, ZHANG Peng, CUI Yixuan

学生：孙志伟，郭浩哲 Students: SUN Zhiwei, GUO Haozhe

建筑设计研究（一）ARCHITECTURAL DESIGN RESEARCH 1

基本设计
BASIC DESIGN

冷 天
LENG Tian

课题内容：校园场所微改造

作为遗产保护领域中的永恒矛盾，保护与利用的互动关系深刻地影响着建筑师对待历史建筑实践的态度。当下中国的城市建设发展，业已从"增量开发"走向"存量更新"。其中，大量历史性建筑（文物建筑、历史建筑、文化遗产等），都面临着如何在保存特有历史文化价值的前提下，充分活化利用其原有空间（内部、外部）的难题。本课程以南京大学鼓楼校区中的历史建筑、场所为操作对象，引导学生理解设计的逻辑性与综合性，综合考虑校园历史文脉、现状格局和未来发展等多方面因素，从创造性的概念构思和建筑学的基本问题出发，对校园中规模较小的建筑、场所等进行深入的改造设计，最终在历史和现实之间取得平衡，通过一个富有创造性的设计来激发既有空间的活力。

设计合作

每个设计工作组由2~3位同学组成，抽签决定设计研究对象。

设计成果

1. 完整的平面图、立面图、剖面图图纸，建议比例不低于1：100，需布置家具以及表达人的行为；
2. 表达空间关系的三维剖透视图，建议比例不低于1：50，要求表达大致构造层次，需布置家具和人的行为表达；
3. 表达设计意图的大样图，建议比例不低于1：20；
4. 表达形体和空间的透视图、轴测图等，可以是渲染图也可以是模型照片；
5. 其他有利于设计意图表达的图纸。

教学进度

第一周：基地调研及场地分析；
第二周：提出功能策划及概念方案可能思路比较；
第三周：概念方案；
第四周：深化设计；
第五周：中期考核；
第六周：绘制图纸；
第七周：设计表达；
第八周：答辩准备。

Subject content: Micro-renovation of campus places
As an eternal contradiction in the field of heritage protection, the interactive relationship between protection and utilization has profoundly affected architects' attitudes towards historical architectural practices. The current urban construction and development in China has shifted from "incremental development" to "stock renewal". A large number of historic buildings (cultural relics, historical buildings, cultural heritage, etc.) are faced with the problem of how to fully

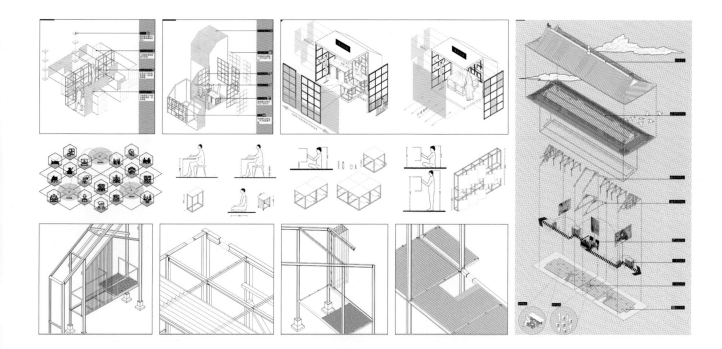

activate and utilize the original spaces (inside and outside) while preserving their unique historical and cultural values. This course takes the historical buildings and places in the Gulou Campus of Nanjing University as the operating object, guides students to understand the logic and comprehensiveness of the design, comprehensively considers the historical context of the campus, the current situation and future development and other factors. Starting from creative concepts and the basic issues of architecture, we will carry out in-depth renovation and design of small-scale buildings and places on the campus, and finally achieve a balance between history and reality, and stimulate the vitality of the existing space through a creative design.

Design cooperation
Each design working group is composed of 2-3 students, and draw lots to determine the design research objects.

Design achievements
1. Complete plan, elevation and section drawings, the recommended scale is not less than 1 : 100, and furniture needs to be arranged and behaviors of people expressed;
2. The three-dimensional cross-sectional perspective view to express the spatial relationship is needed, the recommended scale is not less than 1 : 50, it is required to express the general structure level, and it is necessary to arrange furniture and express human behaviors;
3. The large-scale drawing to express the design intention is required, the recommended scale is not less than 1 : 20;
4. Perspective drawings, axonometric drawings, etc. to express the shape and space are required, which can be renderings or model photos;
5. Other drawings that are conducive to the expression of design intention.

Teaching progress
Week 1: Base survey and site analysis;
Week 2: Functional planning and comparison of possible ideas for conceptual schemes proposal;
Week 3: Concept project;
Week 4: Deepening design;
Week 5: Mid-term assessment;
Week 6: Drawings;
Week 7: Design expression;
Week 8: Defense preparation.

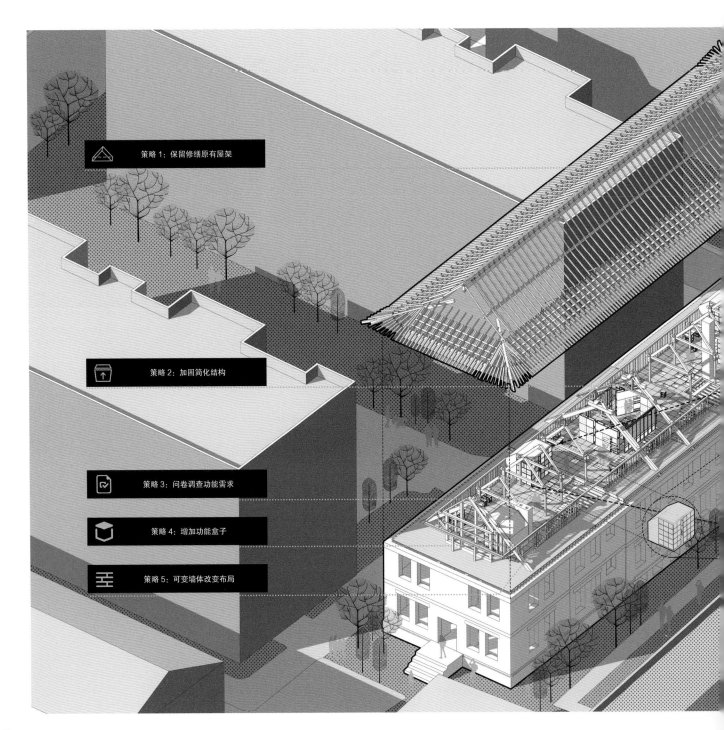

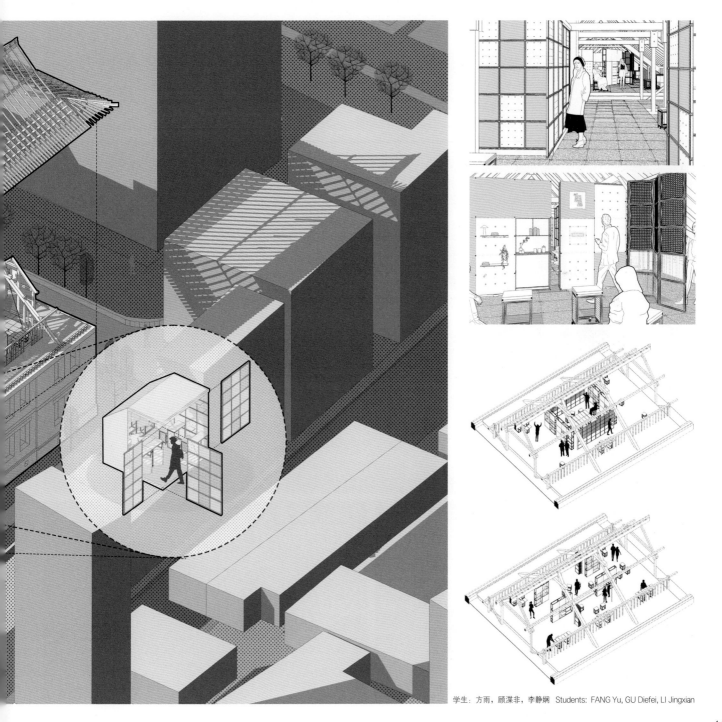

学生：方雨，顾渫非，李静娴　Students: FANG Yu, GU Diefei, LI Jingxian

学生：陈露茜，杨佳锟 Students: CHEN Luxi, YANG Jiakun

建筑设计研究（一）ARCHITECTURAL DESIGN RESEARCH 1

概念设计
CONCEPTUAL DESIGN

鲁安东
LU Andong

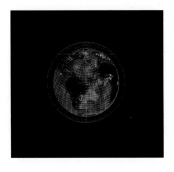

研究对象：作为记忆介质的建筑学

长久以来，建筑在城市中承担着集体记忆的介质功能。而在当代，数字技术在日常生活中的普遍运用，从根本上改变了城市记忆的塑造形式和接受形式。建筑作为城市记忆的枢纽介质，必须整合当代城市记忆的技术架构。随着"记忆的艺术"逐渐被更替为"记忆的技术"，这也带来了全新的创造可能，同时也要求对"设计"这一工作进行重新定义。此次课程将从城市记忆的空间介质入手，对城市轴线、物体性、景象／图像、叙事、空间仪轨等中介机制进行研究，进而提出一个针对性的设计计划。

课程要求

学生人数：18人，三人一组。
作业要求：各组自定研究选题，并针对选题拟定任务书，完成相关技术图纸绘制。
每组完成概念手册1份（80~100页）。

教学进度

第一周：9.22—9.23；
第二周：9.28—9.29，选题汇报及概念分析；
第三周：国庆；
第四周：9.11—9.12，概念分析汇报及拟定任务书；
第五周：10.19—10.20，中介载体设计（建筑/城市/空间设计）；
第六周：10.26—11.27，中期汇报（外请评委）；
第七周：11.2—11.3，图纸深化；
第八周：11.9—11.10，最终评图。

Research object: Architecture as the medium of memory

For a long time, architecture has assumed the media function of collective memory in the city. In contemporary times, the widespread use of digital technology in daily life has fundamentally changed the molding form and acceptance form of urban memory. As the pivotal medium of urban memory, architecture must integrate the technical framework of contemporary urban memory. As the "art of memory" is gradually replaced by "technology of memory", it also brings new

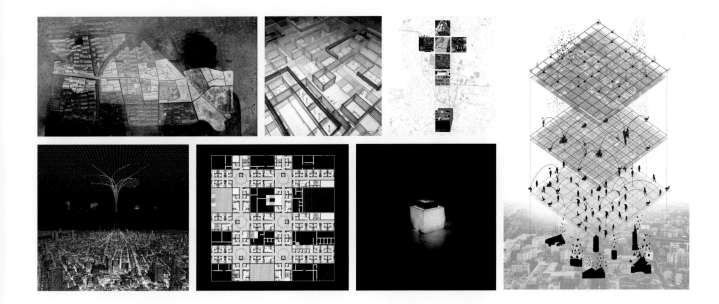

possibilities for creation, and at the same time requires a redefinition of the work of "design". This course will start with the spatial medium of urban memory, research on intermediary mechanisms such as the urban axis, objectivity, scene/image, narrative, space rituals, etc., and then propose a targeted design plan.

Course requirements
Number of students: 18, each group of three.
Homework requirements: Each group decides the research topic by students, draws up a brief for the selected topic, and completes the drawing of relevant technical drawings.
Each group completes a concept booklet (80–100 pages).

Teaching progress
Week 1: 9.22–9.23;
Week 2: 9.28–9.29, topic selection report and conceptual analysis;
Week 3: National Day;
Week 4: 9.11–9.12, conceptual analysis report and task book drafting;
Week 5: 10.19–10.20, intermediary carrier design (architecture/urban/spatial design);
Week 6: 10.26–11.27, mid-term report (outside judges invited);
Week 7: 11.2–11.3, drawing deepening;
Week 8: 11.9–11.10, final review.

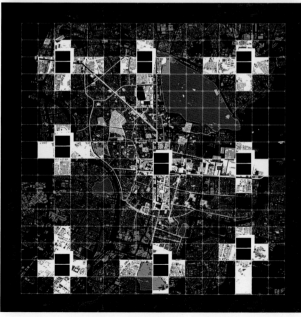
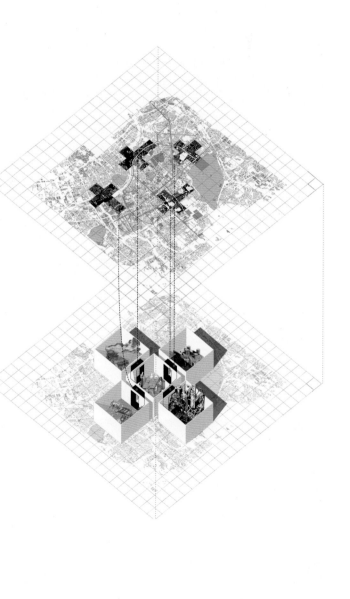

学生:范玉斌,李宇翔 Students: FAN Yubin, LI Yuxiang

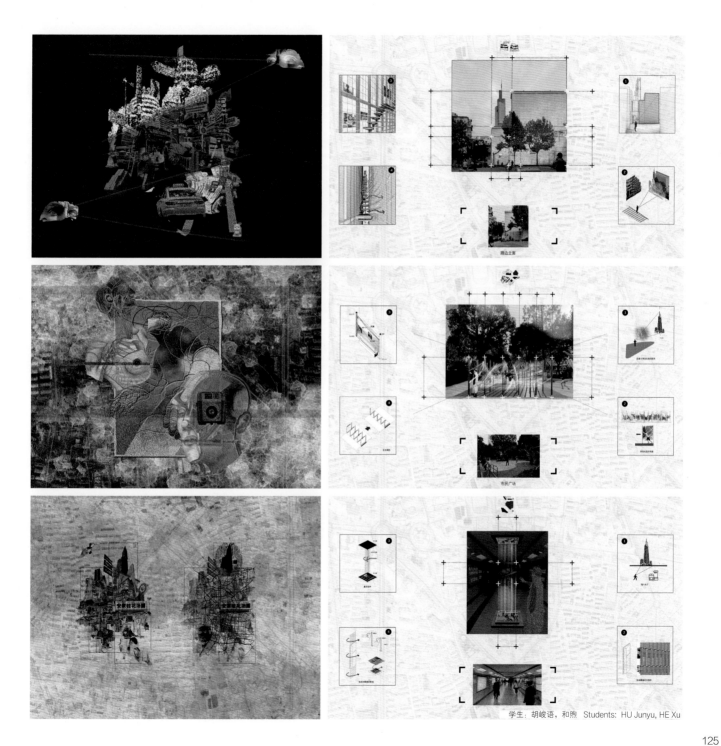

学生：胡峻语，和煦　Students: HU Junyu, HE Xu

建筑设计研究（一）ARCHITECTURAL DESIGN RESEARCH 1

概念设计
CONCEPTUAL DESIGN

周渐佳
ZHOU Jianjia

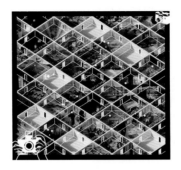

课程内容

2020年的硕士生概念设计受线上空间的启发，以"从物理空间到线上空间"为主题提取空间原型加以设计。线上空间曾经被认为是物理空间的附属，却在很短的时间内成为所有活动发生的重要乃至唯一载体。这个过程也将线上空间推到了建筑学科的面前，我们会发现线上空间是与物理空间并行，且有着同等意义的一个新领域，此前却极少获得来建筑学科的关注。2021年"元宇宙"概念对所有行业形成了冲击，也使得更多可能性展现在我们面前。顺延这条脉络的讨论正当其时。本学期的硕士生概念设计以"从线下到线上"为主题，形成一个开放的讨论过程。课程以建筑系学科最熟悉的"studio"为基本的空间类型，在八周的设计、理论教学中围绕与之相关的线下、线上的概念、行动等展开探究。一方面提供相应的技术支持，另一方面通过系列讲座进一步打开思考的广度与深度。最终成果同样形成展览，在线上与线下空间同时展示。课程中的所有讲座、小论文、讨论、手稿将汇编成册，作为对课程的记录。

讲座

AR、VR与XR ｜ Kyle张彦超，彼真科技创始人，交互技术专家；
虚拟空间补完计划 ｜ 王祥博士，同济大学建筑与城市规划学院讲师；
交互技术 ｜ 万华生态；
想象的空间与想象的维度 ｜ 王令杰，青年艺术家；
影像叙事 ｜ 何伊宁，青年影像艺术策展人；
对等式网络和线上元空间 ｜ 龙星如，策展人、写作者。

课程安排

课程以三次评图为节点，划分成三个阶段。在每两次集中评图之间，围绕着一个特定主题，例如概念、工具、技术手段等，以类似工作坊的方式进行密集工作。
课程采用单人或双人合作的方式，视选课人数而定。

Course contents

Inspired by the online space, the conceptual design of the master students in 2020 is based on the theme of "from physical space to online space" to extract the space prototype and design. The online space was once considered an adjunct to the physical space, but it has become an important and even the only carrier for all activities in a short period of time. This process also pushes the online space to the front of the architectural discipline. We will find that the

online space is a new field that is parallel to the physical space and has the same meaning. It has received little attention from the discipline before. In 2021, the concept of "Metaverse" had an impact on all industries, and more possibilities will be presented to us. It is timely to discuss on the extension of this thread. The concept design of master students in this semester takes "from offline to online" as the theme, forming an open discussion process. The course takes the "studio" which is the most familiar to architecture students as the basic space type, and explores related offline and online concepts and actions during the eight-week design and theoretical teaching. On the one hand, it provides corresponding technical support, on the other hand, it further opens up the breadth and depth of thinking through a series of lectures. The final results also form an exhibition, which is displayed both online and offline. All lectures, essays, discussions, and manuscripts in the course will be compiled into a book as a record of the course.

Lectures
AR, VR and XR | Kyle Zhang Yanchao, founder of Bizhen Technology, interactive technology expert;
Complementary plan for virtual space | Dr. Wang Xiang, Lecturer, School of Architecture and Urban Planning, Tongji University;
Interactive technology | Wanhua Ecology;
Imaginary space and imaginary dimension | Wang Lingjie, young artist;
Video narrative | He Yining, curator of young video art;
Peer-to-peer network and online Metaspace | Long Xingru, curator and writer.

Course arrangement
The course is divided into three phases with three reviews as nodes. In between each of the two reviews, intensive work is carried out in a studio way around a specific theme, such as concepts, tools, technical means, etc..
The course is conducted in one or two students depending on the number of participants.

学生：钟言，陈宜旻　Students: ZHONG Yan, CHEN Yimin

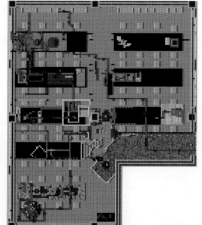

学生：程科懿，闫朝新　Students: CHENG Keyi, YAN Chaoxin

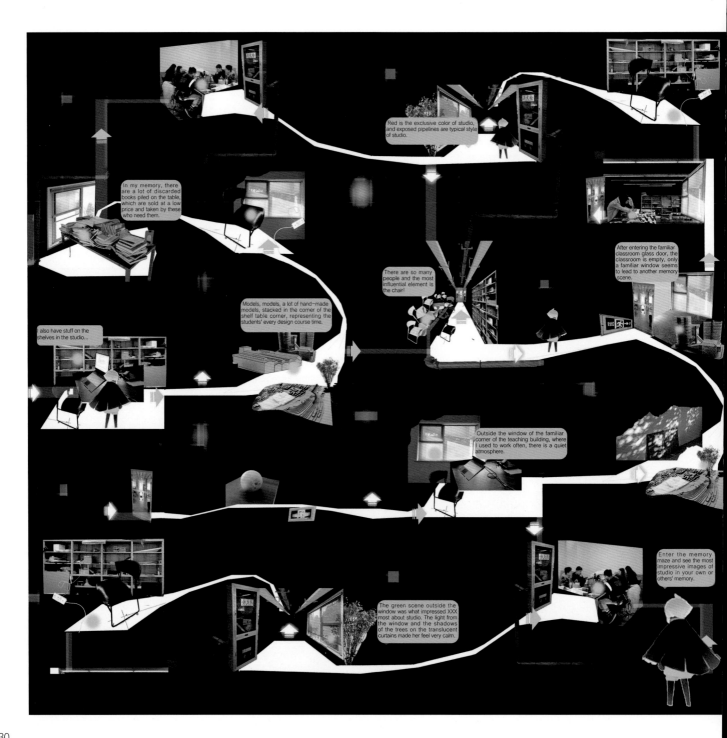

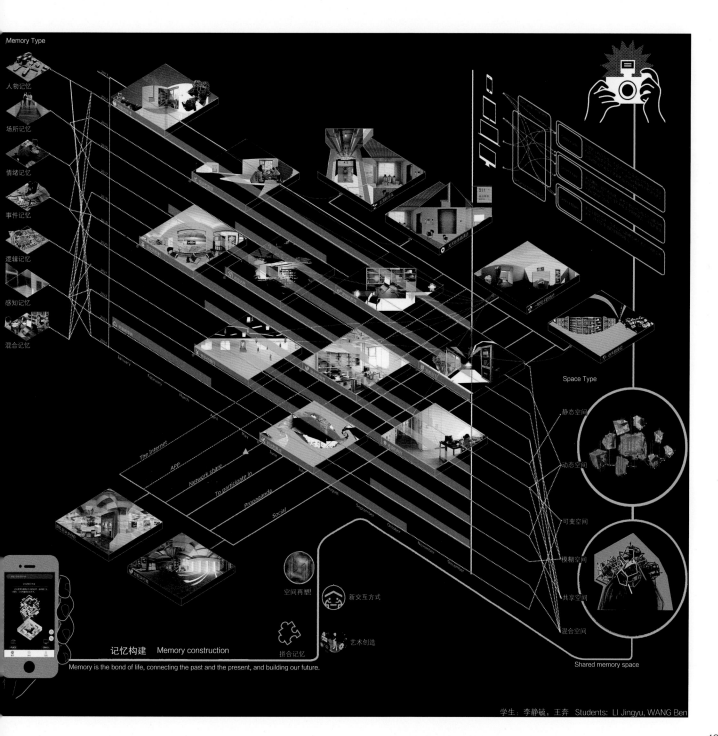

建筑设计研究（二）ARCHITECTURAL DESIGN RESEARCH 2

综合设计
COMPREHENSIVE DESIGN

程 向 阳
CHENG Xiangyang

课程任务
  中国科学院南京土壤研究所位于南京市玄武区北京东路71号，成立于1953年，其前身为1930年创立的中央地质调查所土壤研究室。中国科学院南京地理与湖泊研究所位于南京市玄武区北京东路73号，其前身系1940年8月成立的中国地理研究所，1988年1月改名为中国科学院南京地理与湖泊研究所并沿用至今，是中国唯一以湖泊—流域系统为主要研究对象的中国地理研究所。上述两个研究所共用一个大院空间，没有明确的物理边界分割，总占地面积34 760 m^2。

设计任务
  在相关调研、分析和上位规划基本条件前提下，在场地范围内规划设计一个以营造科研、教育、文化或社区配套等为主体内容的开放性城市公共空间场所。由设计者研究自定主体内容以及其他兼容业态，如休闲娱乐、酒店、商业、商务以及公寓等。
  规划设计除两个历史建筑必须保留外，应最大化利用现有建筑空间价值，尊重现有重要的空间脉络，设计者可结合规划设计，自定未来改造用途以及改造力度。
  采取小组合作形式。

成果要求
1. 城市设计总平面图（1∶1 000）与总体表现图；
2. 相关研究、策划、目标及概念提出；
3. 场地与建筑空间诸系统分析图解；
4. 重点项目建筑、场地环境深化设计（1∶300~1∶500）；
5. 表达设计意图的表现图和实体模型（比例待定）；
6. A3文本1套，PPT成果展示，6张A0图纸。

课程安排
阶段1：前期研究／第一至二周；
阶段2：概念结构规划／第三至四周；
阶段3：重点项目建筑和场地环境深化设计／第五至七周；
阶段4：课程答辩／第八周。

Course tasks
The Institute of Soil Science, Chinese Academy of Sciences is located at No. 71, Beijing East Road, Xuanwu District, Nanjing. The Institute of Soil Research was established in 1953, which was formerly known as the Soil Research Office of the Central Geological Survey Institute found in 1930. The Nanjing Institute of

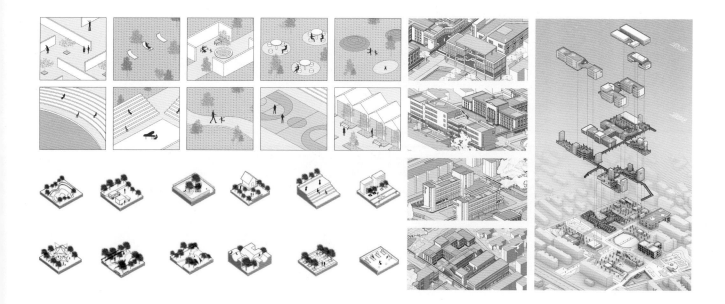

Geography and Limnology of the Chinese Academy of Sciences is located at No. 73, Beijing East Road, Xuanwu District, Nanjing, which was formerly known as the Chinese Institute of Geography, found in 1940, August. It was renamed the Institute of Earth and Lakes is still in use today. It is the only Chinese institute of Geography that focuses on lake-watershed systems as its main research object. The above-mentioned two research institutes share a compound space without clear physical boundary division, covering a total area of 34 760 m^2.

Design tasks
Under the premise of relevant research, analysis and upper-level planning, plan and design an open urban public space within the scope of the site, with the main content of scientific research, education, culture or community facilities. The designer studies and defines main content and other compatible formats, such as leisure and entertainment, hotels, commerce, business and apartments, etc..
In addition to the two historical buildings that must be preserved, the planning and design should maximize the use of the existing architectural space and respect the existing important space context. The designer can combine the planning and design to determine the future renovation use and renovation intensity.
Adopting a group collaboration format.

Achievement requirements
1. Urban design general plan (1 : 1 000) and overall performance map;
2. Relevant research, planning, the proposal of goal and concept;
3. Diagrams for analysis of various systems of site and architectural space;
4. In-depth design of key project buildings and site environment (1 : 300-1 : 500);
5. The performance diagram and physical model expressing the design intention (scale to be determined);
6. A set of A3 texts, display of PPT results, 6 sheets of A0.

Course arrangement
Phase 1: Preliminary Research / week 1-2;
Phase 2: Conceptual structure planning / week 3-4;
Phase 3: Deepening design of architecture and site environment of key projects / week 5-7;
Phase 4: Course defense / week 8.

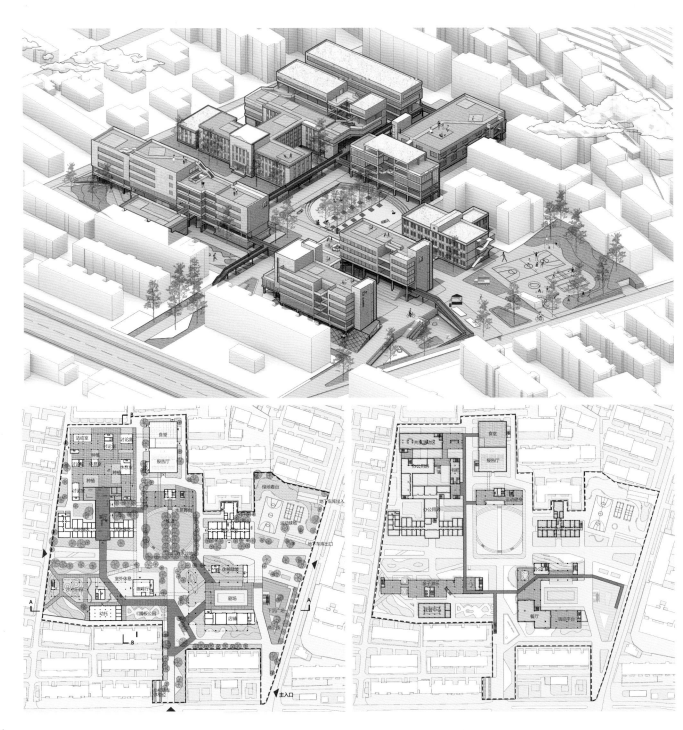

学生：骆婧雯，孙志伟，晏攀，周盟珊，周航　Students: LUO Jingwen, SUN Zhiwei, YAN Pan, ZHOU Mengshan, ZHOU Hang

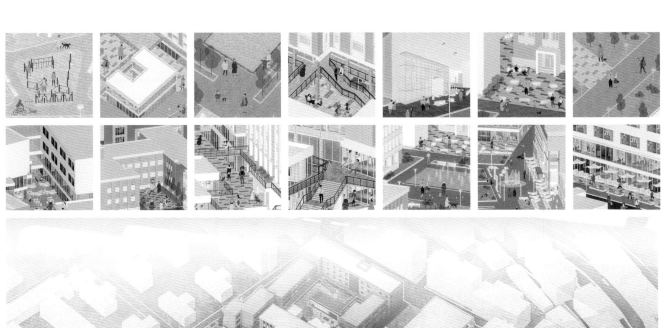

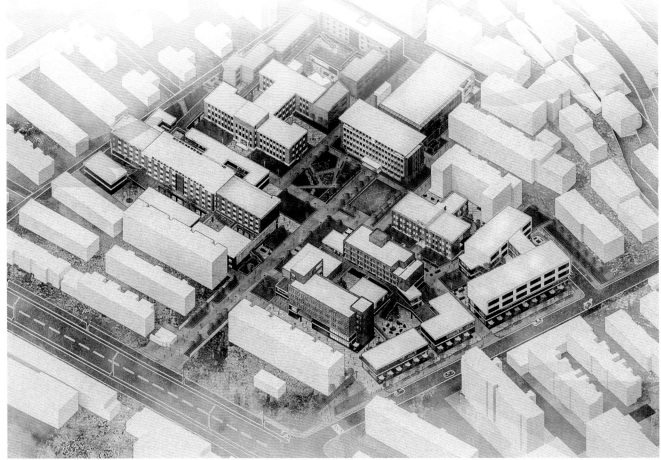

学生:杨晟铨,徐一凡,杨子征,邵鑫露 Students: YANG Shengquan, XU Yifan, YANG Zizheng, SHAO Xinlu

学生:李鹿,孙珂,谢宇航,张鹏 Students: LI Lu, SUN Ke, XIE Yuhang, ZHANG Peng

建筑设计研究（二）ARCHITECTURAL DESIGN RESEARCH 2

综合设计
COMPREHENVESIVE DESIGN

金 鑫
JIN Xin

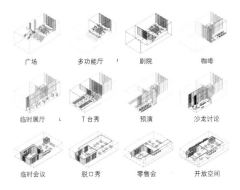

广场　多功能厅　剧院　咖啡

临时展厅　T台秀　预演　沙龙讨论

临时会议　脱口秀　零售会　开放空间

研究课题

城市工业建筑空间再生研究——南京国家领军人才创业园28号楼（原第二机床厂大厂房）改造设计

研究问题

从社会发展的规律性来看，由于工业文明社会向后工业文明社会的转型与过渡，部分工业建筑必然面临停产、搬迁、转移、废弃等境况。而大规模的拆除、重建，必然产生极大的资源浪费、污染排放等问题。因此，以空间再生为基本特征的工业建筑再利用，成为建筑师的新任务。本次设计研究将聚焦工业建筑空间的再利用问题，根据工业建筑空间的特点，结合社会需求的新内容和空间要求，努力探索工业建筑空间再生的规律。

1. 尺度转换

工业建筑、构筑物和场地等通常具有远超人体尺度的巨大体量，并容纳大量复杂的机器和设备的运作。这样的物质环境，具有机器化、非人性的尺度和空间，难以与常人的生活、工作等活动相关联。以空间的再生作为设计研究的核心，意味着将工业建筑的巨大空间转向民用、公共的空间。

而工业建筑巨大的空间尺度和坚固的结构体系，提供了重新组织交通流线的可能。这同样需要作相应的空间尺度转换研究，以汽车和车行的尺度作为基本单元来研究工业建筑的空间适应性。

2. 程序重置

在原为满足生产工艺流程要求而设置的工业建筑空间及其组织关系中，重新置入符合城市生活需求的新的程序与功能，合理安排新的活动内容。

3. 结构重组

在工业建筑的改造中，为了满足空间再利用的需求，可对工业建筑既有结构体系进行改变或重组。新置入的结构体与既有工业建筑结构体系可能形成多种空间位置和受力关系。

采取小组合作形式。

课程信息

指导教师：金鑫。
合作指导：杨侃。
学生人数：21人，三人一组；
　　　　结合课程内容进行课堂大组讨论、小组讨论和讲课。
作业要求：自拟任务书、模型、A0图纸、汇报PPT等。
系列讲座拟请嘉宾：
　　　周苏宁，米思建筑；
　　　王子耕，Pills工作室；
　　　马岛，来建筑；
　　　罗宇杰，罗宇杰工作室。

Research topic

Research on Urban Industrial Building Space Regeneration—Renovation Design of Building 28 of Nanjing National Leading Talents Pioneering Park (formerly the large workshop of the Second Machine Tool Factory)

Research problems

From the perspective of the regularity of social development, due to the transformation and transition from an industrial civilized society to a post-

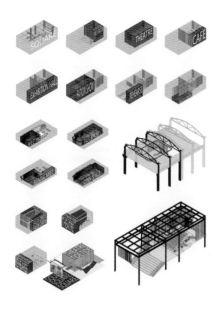

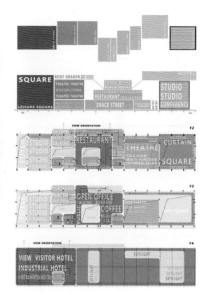

industrial civilized society, some industrial buildings will inevitably face situations such as suspension of production, relocation, transfer, and abandonment. However, large-scale demolition and reconstruction will inevitably lead to huge waste of resources, pollution discharge and other problems. Therefore, the reuse of industrial buildings characterized by space regeneration has become a new task for architects. This design research will focus on the reuse of industrial building space. According to the characteristics of industrial building space, combined with the new content and space requirements of social needs, we will strive to explore the law of industrial building space regeneration.

1. Scale conversion

Industrial buildings, structures, and sites usually have huge volumes that far exceed the scale of the human body, and accommodate the operation of a large number of complex machines and equipment. Such a physical environment has a mechanized and inhuman scale and space, and is difficult to relate to ordinary people's life, work and other activities. Taking the regeneration of space as the core of design research means turning the huge space of industrial buildings into civil and public spaces.

The huge spatial scale and solid structural system of industrial buildings provide the possibility to reorganize the traffic flow. This also requires corresponding research on spatial scale conversion, using the scale of cars and car dealerships as the basic unit to study the spatial adaptability of industrial buildings.

2. Program reset

In the industrial building space and its relationship that are originally set up to meet the requirements of the production process, new procedures and functions that meet the needs of urban life are re-installed, and new activities are reasonably arranged.

3. Restructuring

In the transformation of industrial buildings, in order to meet the needs of space reuse, the existing structural system of industrial buildings can be changed or reorganized. The newly placed structure may form a variety of spatial positions and stress relationships with the existing industrial building structure system. Adopting a group collaboration format.

Course informations

Instructor: Jin Xin.
Co-director: Yang Kan.
Number of students: 21, each group of three;
   Classroom group discussions, small group discussions and lectures are conducted in conjunction with the course content.
Homework requirements: Self-created briefs, model, A0 drawing, report PPT, etc..
Invited guests for the lecture series:
   Zhou Suning, Misi Architecture;
   Wang Zigeng, Pills Studio;
   Maldives, Atelier LAI;
   Luo Yujie, Luo Yujie Studio.

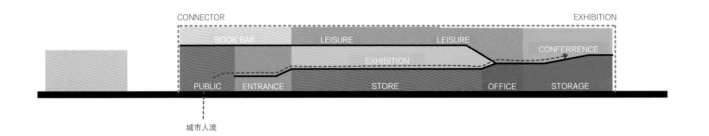
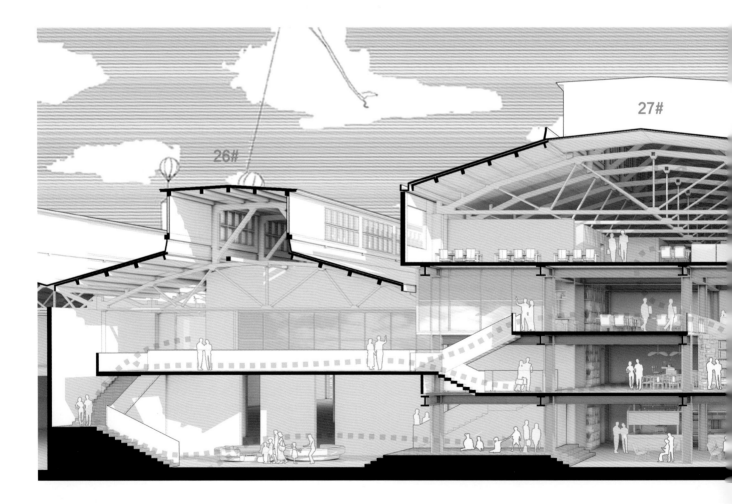

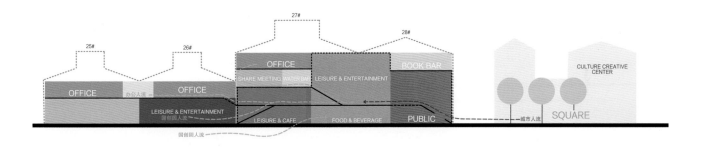

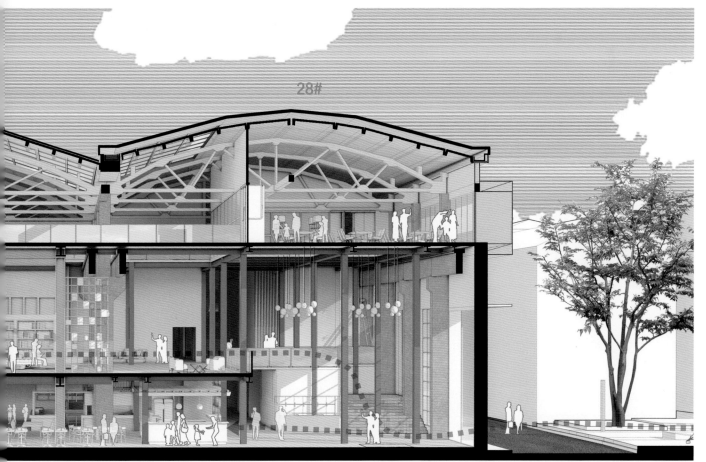

学生：邱国强，祝诗雅，翟曌钰，冯子恺　Students: QIU Guoqiang, ZHU Shiya, ZHAI Zhaojue, FENG Zikai

学生:王淇泓,钟子超,刘玥蓉 Students: WANG Qihong, ZHONG Zichao, LIU Yuerong

建筑设计研究（二）ARCHITECTURAL DESIGN RESEARCH 2
城市设计
URBAN DESIGN

华晓宁
HUA Xiaoning

研究课题
多重文脉中的基础设施都市化——南京中华门火车站地段再生

课程议题
基础设施是当代城市研究与实践的重要主题。作为为社会生产和居民生活提供公共服务的物质工程设施及其系统，它保障着城市有机体的运行，同时又是城市物质空间系统的重要组成部分，自身便占据了场址，界定了空间，形成了场所，连接成系统，构筑了场域。

以往基础设施仅仅被视作市政工程的专业领域，遵循工具理性。被传统建筑学忽视多年，许多基础设施成为城市中消极和被动的要素。随着城市的演进与变迁，既往的基础设施场址面临着再生。然而，在漫长的既往，基础设施早已沉浸在多重的城市文脉中。如何在复杂的城市多重文脉中重定义基础设施，将其视为重要的城市操作性对象与媒介，对基础设施"赋能"并转化为城市中更为积极、能动的场所，激发城市活力，是本课题的主要目标。

对象与场址
民国二十四年（1935年），中华门站随江南铁路（宁芜线以及皖赣铁路芜湖至宣城孙家埠段的前身）的建成而同步投入使用。中华门站始建时有两栋复合式连体平房，为江南铁路公司的张静江总裁斥资聘用上海营造厂建造，站房一直保留使用至今，月台为2台设计，从站房侧数起，一道为正线，二、三、四道为侧线。一站台的雨棚为张静江亲自设计，铁柱和黄色石棉瓦覆盖，并以电灯照明。2014年10月14日，中华门站停止客运业务。

中华门车站是南京老城与新城区衔接的重要节点，周边的城市文脉复杂，历史、自然、人文要素相互交织。

课程要求
对中华门火车站及其周边地段（东至雨花路，西至虹悦城，南到雨花西路、雨花东路，北至应天高架）进行深入调研，分析存在问题与矛盾，构想该区段未来愿景，自行拟定任务书，提出改造更新策略，完成方案设计。方案需将城市、建筑、环境景观综合考虑，进行整合设计。

Research topic
Infrastructure Urbanism in the Multi-Context—Regeneration of Zhonghuameng Railway Station area in Nanjing

Course topic
Infrastructure is an important topic in contemporary urban research and practice. As a material engineering facility and its system to provide public services for social production and residents' life, it guarantees the operation of the urban organism, and at the same time is an important part of the urban material space

system, which occupies the site, defines the space, forms the place, connects into a system, and constructs the field.

In the past, infrastructure was regarded only as a professional field of municipal engineering, following instrumental rationality. Neglected by traditional architecture for many years, much infrastructure becomes a passive and negative element in the city. With the evolution and change of the city, the former infrastructure sites are facing regeneration. However, over the long past, infrastructure has been immersed in multiple urban contexts. How to redefine infrastructure in complex urban contexts, regard it as an important urban operational object and medium, "empower" infrastructure and transform it into a more active and dynamic place in the city, and stimulate the vitality of the city, is the main goal of this course.

Object and site

In the 24th year of the Republic of China (1935), Zhonghuamen Station was put into use simultaneously with the completion of the Jiangnan Railway (the predecessor of the Ningwu Line and the Wuhu-Xuancheng Sunjiabu section of the Anhui-Gan Railway). When Zhonghuamen Station was built, there were two composite joined bungalows. Zhang Jingjiang, president of Jiangnan Railway Company, paid a huge price and hired Shanghai Construction Factory to build them. The station building has been retained and used to this day, and the platform is designed for 2 units, starting from the side of the station building, the first line is the main line, and the second, third and fourth are the side lines. The canopy of platform one was designed by Zhang Jingjiang himself, covered with iron pillars and yellow asbestos tiles, and illuminated by electric lights. On October 14, 2014, passenger services ceased at Zhonghuamen Station.

Zhonghuamen Station is an important node connecting Nanjing's old town and new town, and the surrounding urban context is complex, and historical, natural and human elements are intertwined.

Course requirements

Conduct in-depth investigation on Zhonghuamen Railway Station and its surrounding areas (east to Yuhua Road, west to Hongyue City, south to Yuhua West Road, Yuhua East Road, north to Yingtian Elevated Road), analyze the existing problems and contradictions, conceive the future vision of the section, formulate the assignment, propose the transformation and renewal strategy, and complete the scheme design. The scheme needs to comprehensively consider the urban, architectural and environmental landscapes and carry out integrated design.

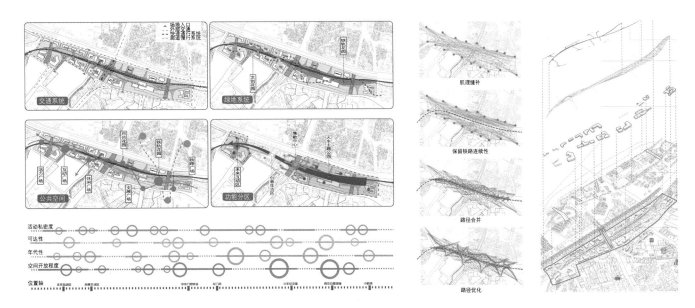
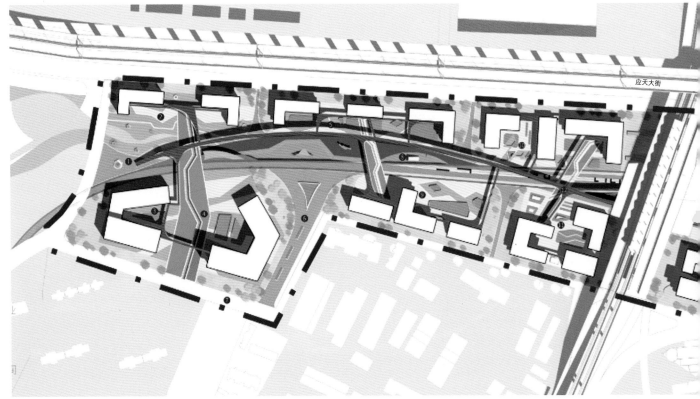

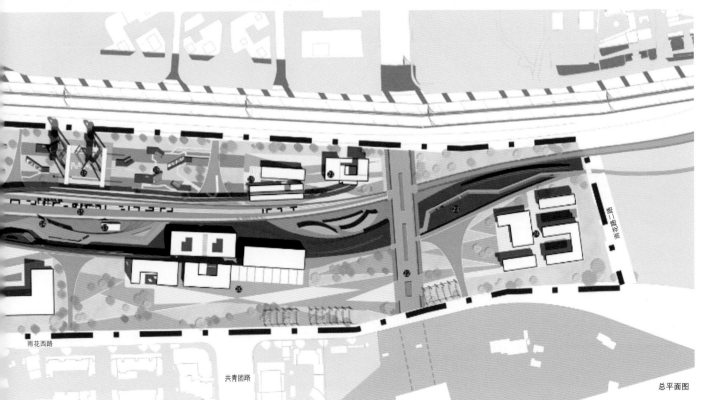

总平面图

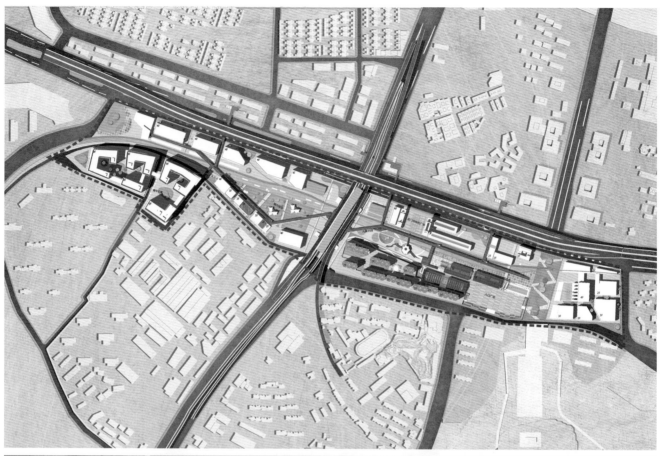

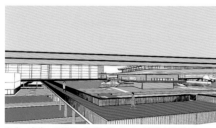
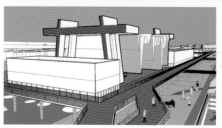

学生：李伟，骆婧雯，陈卓　Students: LI Wei, LUO Jingwen, CHEN Zhuo

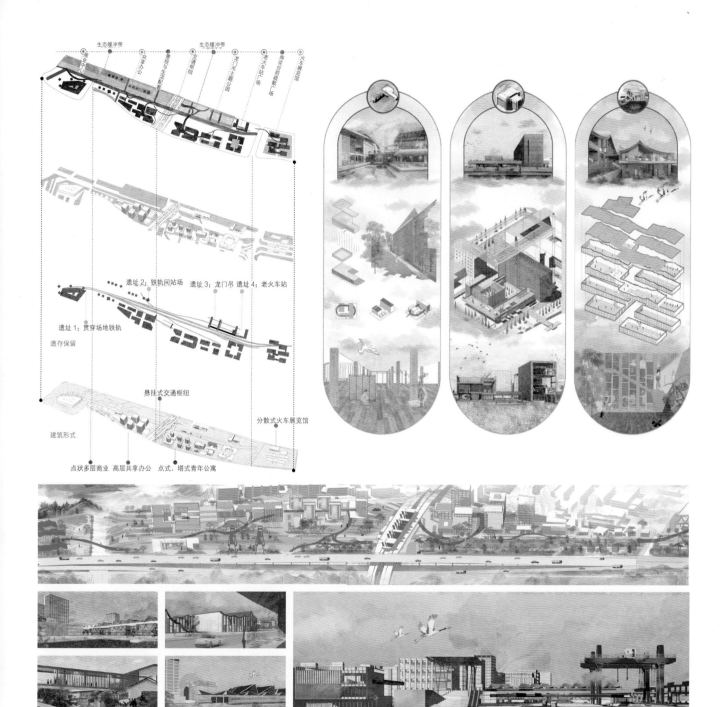

学生：李鹿，邱向楠，赵越 Students: LI Lu, QIU Xiangnan, ZHAO Yue

建筑设计研究（二）ARCHITECTURAL DESIGN RESEARCH 2
城市设计
URBAN DESIGN
胡友培
HU Youpei

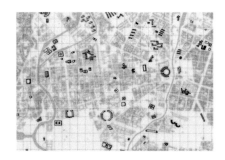

研究课题
　　都市区群岛计划

问题与任务
　　长三角地区是中国城市化程度最高的区域之一，构成了巨大的城市群落与延绵的都市区。在这里，格网城市化模式以巨大的吞吐量吞噬着千百年逐渐形成的人居与自然景观，不断占领多样与差异化的地表，塑造出一座座类似的新城、居住点。都市区中的生活被标准化在格网的城市中，单调而空洞。
　　很大程度上，我们仍然沿用着现代主义城市规划的草纸策略，无差别地对待着都市区复杂的自然与村庄系统。在可持续、生态理念已成为普遍共识的今天，这个草纸的本底似乎并未有多大改观。尽管格网的城市化是一种高效的、低成本的空间生产技术，但其弊端也同样显著并且显得陈词滥调。可能在格网模型化之外，在中心城区外围广袤的都市区中，能够想象另一种城市化模型及其建筑学？

自然的架构
　　在都市区地表上规划设计新的城市区域的首要工作，是遵循并厘清自然的架构。它将构成设计与建设的边界，是都市区大尺度图底关系中的基底，是一种绿色的"空"。

村庄的品质
　　我们对村庄的迷恋与欣赏，不是乡愁主义与旅游主义的。我们欣赏都市区中村庄与自然建立的和谐关系，以及村庄使用简单类型创造的丰富性与在地性。这正是格网城市所欠缺的。

密度
　　密度是城市得以成立的基础，都市区的城市化模型也不例外。否则将继续是村庄，或沦为无边无际的蔓延带。我们将保持与中心城区类似的毛密度，以在城市运行层面，确保城市化的合理性。

TOD与效率
　　依托当代城市高效的交通系统，地铁、轻轨、高速路，使得我们今天有可能在广袤的都市区中，一方面享受与自然的近亲与大量的绿色开放空间，同时又保持与城市母体紧密连接。TOD，交通导向的发展模式，是我们构想新城市化模式的基础设施保障。

城市的尺度
　　除了密度外，村庄和城市在物质形态层面的区别是什么？一个显而易见的区别在于城市具有村庄所不具备的大尺度的人造物，或某种城市结构。在古典时代，这也许是巨大的纪念物；在巴洛克时代，甚至延续至今，则是巨大的轴线；在先锋派那里，发明了巨构，是群形式；在昂格斯那里，是大形；到了雷姆·库哈斯，则成为关于"大"的建筑学。无论如何，我们在这些巨大的人造物中看到了村庄所不具备的集体性、都市性，一种人类社会在个体多样性基础之上建立的空间秩序与对集体意识的呈现、表达。这是村庄无法获得的品质，也是城市永恒的魅力所在。

Research topic
Archipelago Plan in Metropolitan Area

Issues and tasks

The Yangtze River Delta region is one of the most urbanized regions in China, constituting a huge urban agglomeration and a stretch of urban areas. Here, the grid urbanization model devours the human settlements and natural landscapes with huge throughput that have been gradually formed over thousands of years, constantly occupying diverse and differentiated surfaces, and shaping similar new cities and settlements. Life in metropolitan areas is standardized in a grid of cities, monotonous and empty.

To a large extent, we still follow the sketch strategy of modernist urban planning, which treats the complex natural and rural systems of urban areas without discrimination. Today, when the concept of sustainability and ecology has become a general consensus, the background of this sketch does not seem to have changed much. Although the urbanization of grids is an efficient, low-cost space production technology, its drawbacks are just as significant yet cliched. Could it be that in addition to grid modeling, in a vast metropolitan area outside the central urban area, imagine another model of urbanization and its structure?

Natural architecture

The first task in planning and designing new urban areas on the surface of an metropolitan area is to follow and clarify the natural structure. It will constitute the boundary between design and construction, and it is the basis in the large-scale map base relationship of the metropolitan area, which is a green "void".

Quality of the village

Our fascination and appreciation of the village is not nostalgic and tourist. We appreciate the harmonious relationship between villages and nature in urban areas, as well as the richness and locality of villages create using simple types. This is exactly what grid cities lack.

Density

Density is the foundation on which cities are founded, and the urbanization model of metropolitan areas is no exception. Otherwise it will continue to be villages, or reduced to the endless sprawl. We will maintain a similar gross density to the central urban area to ensure the rationality of urbanization at the operational level of the city.

TOD and efficiency

Relying on the efficient transportation system of contemporary cities, such as subways, light rails, and highways, it is possible for us to enjoy the close relationship with nature and a large number of green open spaces in the vast urban area today, while maintaining close connection with the urban matrix. TOD, the transit-oriented development model, is the infrastructure guarantee for us to conceive a new urbanization model.

The scale of the city

Apart from density, what is the difference between villages and cities at the level of the physical form? One obvious difference is that cities have large-scale artifacts, or some kind of urban structure, that villages do not have. In classical times, this was perhaps a monument; In the Baroque era, and even to this day, it was a huge axis; In the avant-garde, the mega-structure was invented, the group form was invented; In the case of Unges, it was the grossform; In Rem Koolhaas, it became architecture about the "big". In any case, we see in these huge man-made objects a collectivity and urbanity that villages do not have, a spatial order and expression of collective consciousness established by human society on the basis of individual diversity and the presentation. This is a quality that cannot be obtained in villages, and it is also the timeless charm of cities.

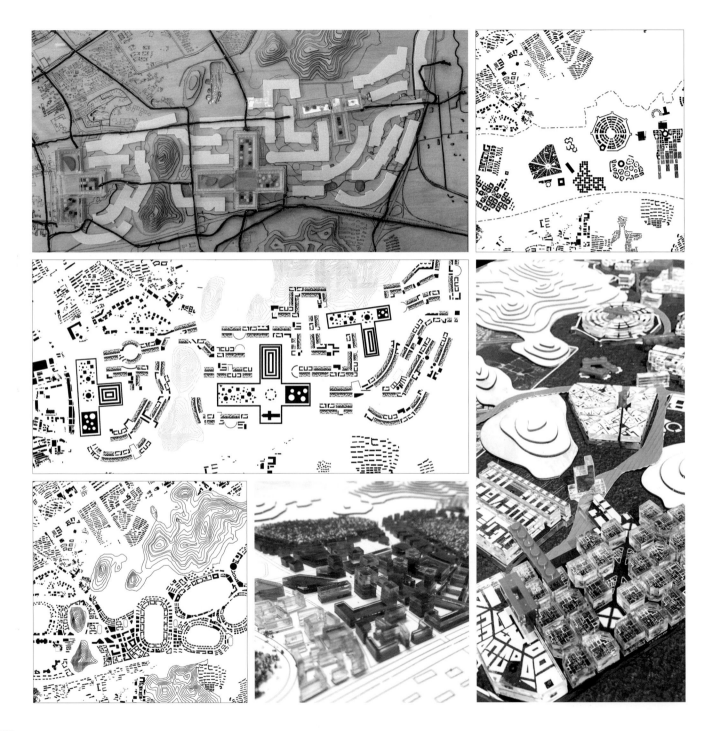

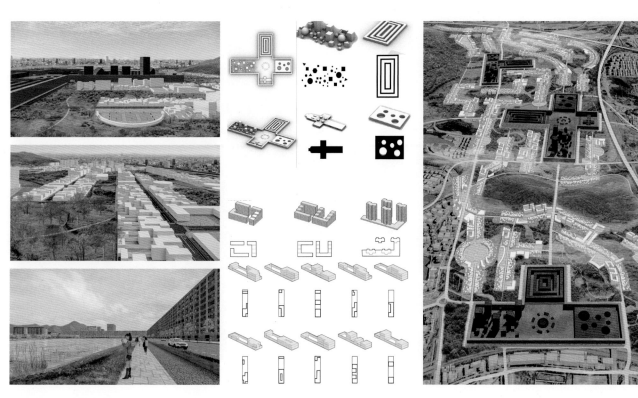

学生：陈露茜，程科懿，邵鑫露，董志昀　Students: CHEN Luxi, CHENG Keyi, SHAO Xinlu, DONG Zhiyun

学生：黄柯，卢卓成，林晴，李文秀　Students: HUANG Ke, LU Zhuocheng, LIN Qing, LI Wenxiu

学生：罗婷，王春磊，闫朝新，朱颜怡　Students: LUO Ting, WANG Chunlei, YAN Chaoxin, ZHU Yanyi

研究生国际教学工作坊 POSTGRADUATE INTERNATIONAL DESIGN STUDIO

光 几 何
LIGHT GEOMETRIES
约瑟夫·施瓦茨
Joseph SCHWARTZ

背景介绍
建筑设计很少将结构性能和太阳能控制视为概念设计阶段的组成部分。事实上，设计概念通常是首先发展的，而结构和太阳能控制方面则在此之后进行分析，其唯一目的是评估设计的性能。此外，结构和太阳能性能的分析是独立进行的，使用不同的模型和设计参数，从而增加了制定互相冲突的设计目标的风险。将结构性能和太阳能控制纳入设计过程有助于降低这种风险。

该工作坊提出了一种新颖的设计策略，旨在将建筑、结构和太阳能控制方面整合为概念设计阶段的组成部分。因此，建筑师和工程师可以在设计过程的早期做出明智的设计决策，以平衡设计的技术和创意方面。基于几何图形的方法，仅考虑几何设计参数，是所提出的整体设计方法的核心，因为它们允许设计人员直接理解单个设计参数对结构和太阳能性能的影响，而不会忽视建筑考虑因素。

教学目标
该工作坊旨在向建筑专业的学生介绍基于几何的结构和太阳能设计工具，并将他们的设计进一步发展为展馆。在工作坊中，学生将使用基于几何图形的方法及其数字化实现来调查建筑、结构和太阳能控制方面。特别是，在参数化数字环境（即 Rhinoceros + Grasshopper）中实施这些方法使学生能够直观而有趣地理解每个设计参数的作用和相互关系。在工作坊使用图形静力学和应力场来设计整体结构行为以及定制的互锁木材到木材连接。此外，用于太阳能控制的图形方法将为设计有效的遮阳设备提供信息。最后，人类行为被纳入这种基于几何形状的设计过程，将结构与特定的人类活动联系起来，从而定义展馆的功能和空间标准。

Background introduction
Architectural design rarely considers structural performance and solar control as integral parts of the conceptual design phase. In fact, the design concept is often developed first, while the structural and solar control aspects are analysed later on with the sole purpose of assessing the design's performance. Moreover, the analyses of structural and solar performance are conducted independently, using different models and design parameters, thus increasing the risk of formulating conflicting design objectives. Incorporating structural performance and solar control into the design process would help mitigate this risk.

This studio proposes a novel design strategy aimed at integrating architectural, structural, and solar control aspects as integral parts of the conceptual design phase. As such, architects and engineers can make informed design decisions early in the design process in an attempt to balance the technical and creative sides of design. Geometry-based graphical methods, which consider only geometric design parameters, are the core of the proposed holistic design approach as they allow the designer to directly comprehend the effect of a single design parameter on structural and solar performance without neglecting architectural considerations.

Teaching objectives
The studio aims to introduce architecture students to geometry-based tools for structural and solar design, and further develop their design into a pavilion. In the studio, students will examine architectural, structural, and solar control aspects using geometry-based graphical methods and their digital implementations. In particular, the implementation of the methods in a parametric digital environment (i.e., Rhinoceros + Grasshopper) allows students to intuitively and playfully understand each design parameter's role and interrelationships. Graphic statics and stress fields are used in the studio to design the global structural behaviour as well as bespoke interlocking timber-to-timber connections. Moreover, graphical methods for solar control will inform the design of effective sun-shading devices. Finally, human behaviour is incorporated into this geometry-based design process to connect structures to specific human activities, thereby defining the pavilion's functional and spatial criteria.

reference point

roof form

    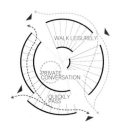

   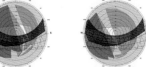

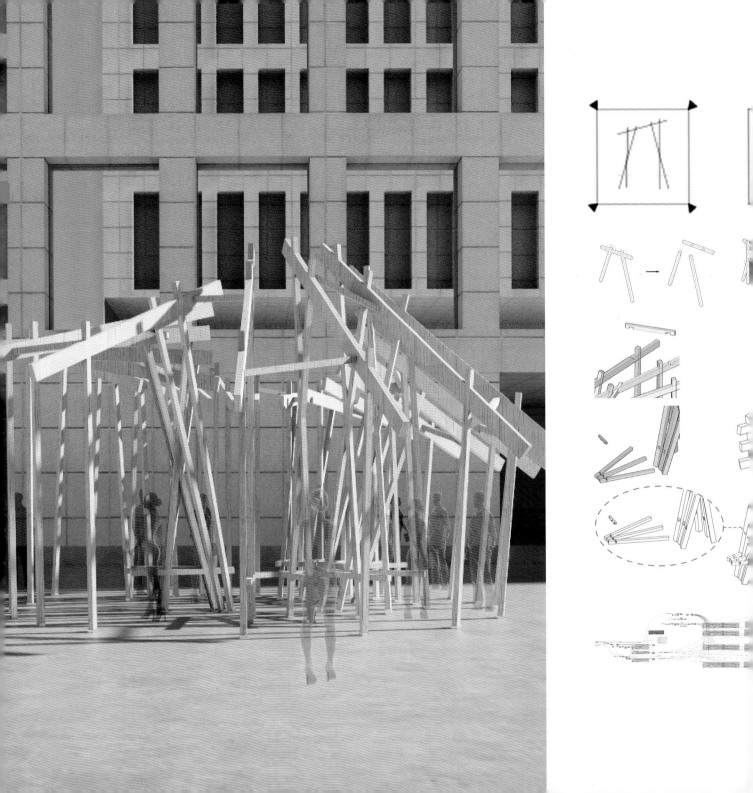

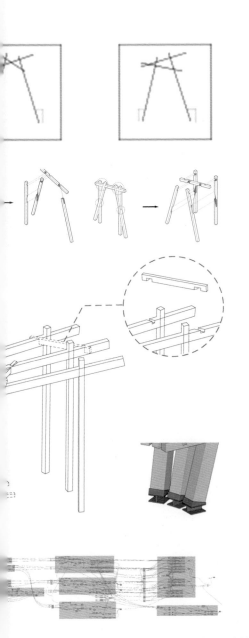

学生：帕·坤奈拉恩西伊，马子昂，赵文，杨茸佳，骆婧雯 Students: Pa PONNAKRAINGSEY, MA Zi'ang, ZHAO Wen, YANG Rongjia, LUO Jingwen

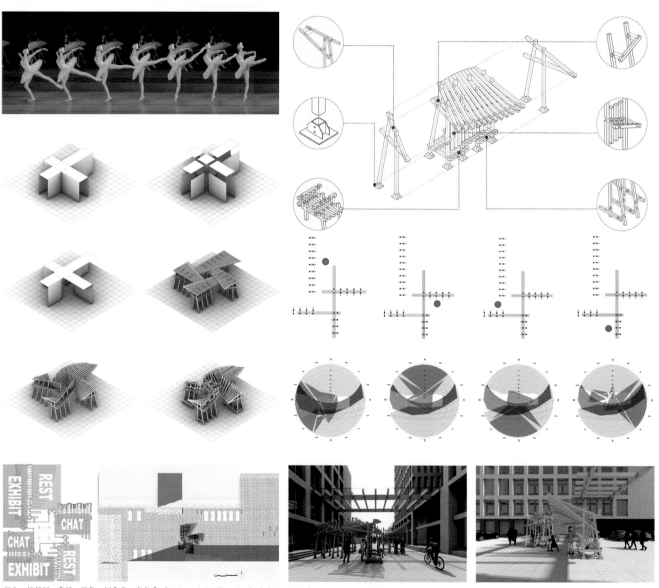

学生：闫朝新，曹超，王鲁，刘鑫睿，卢卓成　Students: YAN Chaoxin, CAO Chao, WANG Lu, LIU Xinrui, LU Zhuocheng

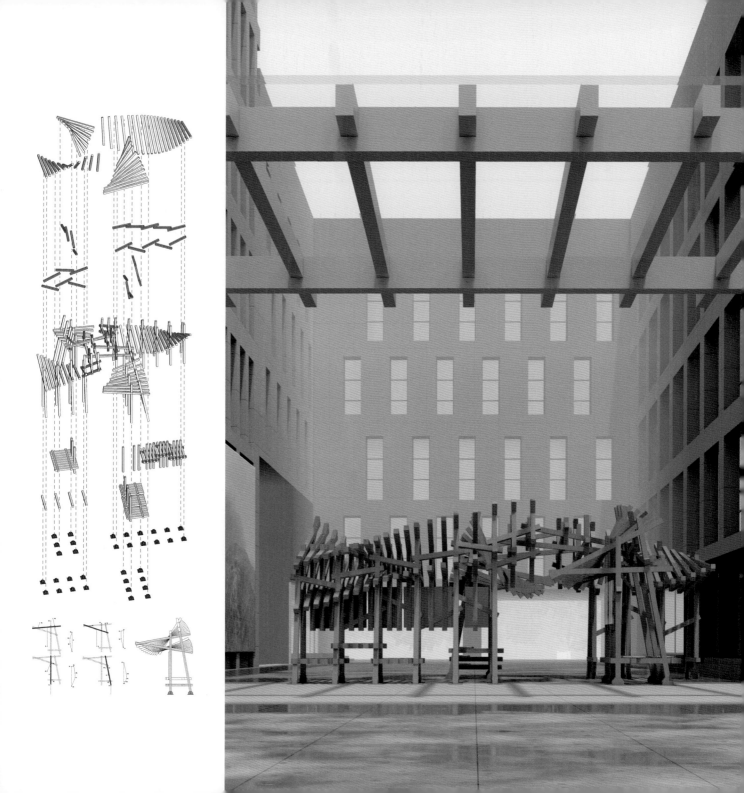

研究生国际教学工作坊 POSTGRADUATE INTERNATIONAL DESIGN STUDIO

可持续的城市密度
SUSTAINABLE URBAN DENSIFICATION

索布鲁赫·赫顿　王洁琼
Sauerbruch HUTTON, WANG Jieqiong

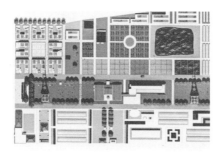

教学目标

在建设城市时，可持续性对我们意味着什么，可持续性会是什么样子，是所有建筑师都要面对的问题。

该工作坊旨在为学生提供方法和操作工具，来探索在提高城市密度、改善生活方式的同时为环境获得更多的效益和减少废物/二氧化碳的产生。技术及其在设计中的应用将作为工作坊的背景知识提供，但答案要求在建筑和城市本身中找到。

课程内容

任务将集中于南京民用无线电短波大楼的开发。

要求构思场地的可持续城市发展，包括现有建筑并使场地致密，同时保持公园，例如它的绿色品质，尽可能与周围环境融为一体。

学生将被要求定义一个程序，并为场地的致密化提出合理的建议，同时结合现有建筑的修复/改建和新建筑的概念。可持续性将成为设计指南，融入城市战略和建筑设计。学生还将被要求考虑使用可回收材料和施工方法。

本次工作坊将支持学生在中国城市/社会背景下探索不同维度和层次的可持续发展的各种表达方式。

第一个主题应该是对"绿色城市"的批判性观点，然后要求学生批判性地研究建筑物的生命周期。首先，他们将被要求审查建筑材料，特别是木材材料的使用。其次，他们需要考虑施工方法。一种方法可能是模块化系统的高级使用，模块化系统在架构方面具有悠久的历史，并且在灵活性、可回收性和生产方面可能是有益的。第三，邀请他们思考建筑的第二次生命——未来的用途变化。

Teaching objectives

What does sustainability mean for us while constructing the city and what would sustainability look like, are the questions for all our architects.

The studio aims to provide students with the methodological and operational tools to densify the city, to improve the lifestyle meanwhile to get more benefit and less waste/ CO_2 production for the environment. Techniques and its application in design would be provided as background knowledge in the studio, but the answer is asked to be found in the architecture and city themselves.

Course contents

The task will focus on developing the Civil Radio Shortwave Building in Nanjing. The task is to conceive a sustainable urban development of the site that includes the existing buildings and densities the site while maintaining the park i.e. green qualities of it, integrating as good as possible in the surroundings.

Students will be requested to define a program and make a reasonable proposal for the densification of the site while combining restoration/conversion of the existing buildings and the conception of new ones. Sustainability would be the design guide, integrated in the urban strategy and architecture design. Students will also be asked to consider the use of recyclable materials and methods of construction.

The studio will support students to explore various expression of sustainability in different dimensions and levels in China urban/society context.

First topic should be the critical view on a "green city". Then the students are requested to critically look into the life cycle of a building. First they would be asked to review construction materials, especially the use of timber. Second, they need to think about the construction method. One method could be the advanced use of modular systems, which have quite a history in structure and could be beneficial in terms of flexibility, recyclability and production. Third, they are invited to think about the second life of a building—the future change of use.

学生：李瑾，卢卓成，杨帆　Students: LI Jin, LU Zhuocheng, YANG Fan

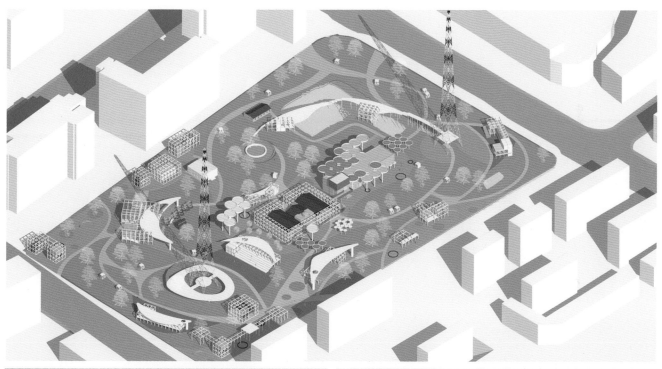
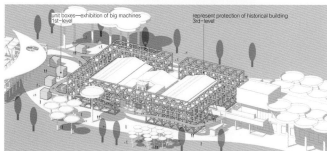
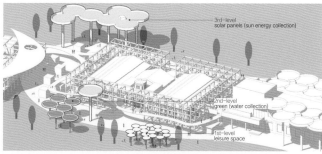
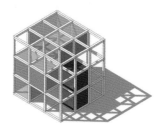
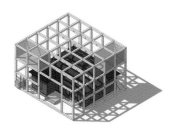

学生：郝欣苑，方雨，张璐　Students: HAO Xinyuan, FANG Yu, ZHANG Lu

建筑设计课程
ARCHITECTURAL DESIGN COURSES

本科一年级
设计基础
· 鲁安东　梁宇舒　刘铨　史文娟　赵潇欣　黄春晓
课程类型：必修
学时学分：64 学时 / 2 学分

Undergraduate Program 1st Year
DESIGN FOUNDATION · LU Andong, LIANG Yushu, LIU Quan, SHI Wenjuan, ZHAO Xiaoxin, HUANG Chunxiao
Type: Required Course
Study Period and Credits: 64 hours/2 credits

课程内容

第一阶段：知觉、再现与设计
知识点：人与物——材料的知觉特征（不同状态下的色彩、纹理、平整度、透光性等）与物理化学特征（成分、质量、力学性能等）；摄影与图片编辑——构图与主题、光影与色彩；三视图与立面图绘制；排版及其工具——标题、字体、内容主次、参考线。

第二阶段：需求与设计
知识点：人与空间——空间与尺度的概念，行为、动作与一个基本空间单元或空间构件尺寸的关系；平面图、剖面图、轴测图的绘制；线型、线宽、图幅、图纸比例、比例尺、指北针、剖断符号、图名等的规范绘制。

第三阶段：制作与设计
知识点：物与空间——建构的概念；空间的支撑、包裹与施工；实体模型制作——简化的建造；计算机建模工具——虚拟建造；透视图绘制。

第四阶段：环境与设计
知识点：人、物与空间——城市形态要素、城市肌理与城市外部空间的概念；街道系统与交通流线；土地划分与功能分类；总平面图、环境分析图（图底关系、交通流线、功能分区、绿地景观系统）、照片融入表达。

Course contents

Phase one: Perception, representation and design
Knowledge points: people and things—the perceptual characteristics of materials (color, texture, flatness, light transmittance, etc. in different states) and physical and chemical characteristics features (composition, quality, mechanical properties, etc.); photography and picture editing—composition and theme, light, shadow and color; three-view and elevation drawings; typography and its tools—headings, fonts, content primary and secondary, and reference lines.
Phase two: Requirements and design
Knowledge points: People and space—the concept of spaces and scale, the relationship between behavior, action and the size of a basic spatial unit or spatial component; plan, section, and axonometric drawings; line types, line widths, map sizes, drawing scales, scale bars, north arrows, section symbols, drawing titles, etc. standardized drawing.
Phase three: Production and design
Knowledge points: Objects and space—the concept of construction; space support, wrapping and construction; physical model making—simplified construction; computer modeling tools—virtual construction; drawing of renderings.
Phase four: Environment and design
Knowledge points: People, objects and space—the concept of urban form elements, urban texture and urban external space; the street system and traffic flow; land division and functional classification; general plan, environmental analysis map (relationship between map and ground, traffic flow, functional zoning, green space landscape system), photos blending into expression.

本科二年级
建筑设计基础
· 刘铨　史文娟
课程类型：必修
学时学分：64 学时 / 4 学分

Undergraduate Program 2nd Year
ARCHITECTURAL DESIGN FOUDATION · LIU Quan, SHI Wenjuan
Type: Required Course
Study Period and Credits: 64 hours/4 credits

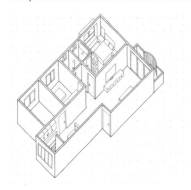

课程内容

本课程是建筑学专业本科生的专业通识基础课程。本课程的任务主要是一方面让新生从专业的角度认知与实体建筑相关的基本知识，如主要建筑构件与材料、基本构造原理、空间尺度、建筑环境等知识，另一方面通过学习运用建筑学的专业表达方法，如平面图、立面图、剖面图、轴测图、实体与计算机模型等来更好地掌握这些建筑基本知识。教学通过认知建筑、认知图示、认知环境等环节建立期学生在这两方面的思维联系，为今后深入的专业学习奠定基础。

Course contents

This course is a professional general education basic course for undergraduate students majoring in architecture. The main task of this course is to enable freshmen to recognize the basic knowledge related to physical architecture from a professional perspective, such as the main building components and materials, basic construction principles, spatial dimensions, building environment and other knowledge. On the other hand, they can better master the basic knowledge of architecture by learning and using professional expression methods of architecture, such as plan, elevation section drawings, axonometric drawing, solid and computer models. Teaching establishes students' thinking connections in cognitive architecture, cognitive graphics, cognitive environment, and other aspects, laying the foundation for future in-depth professional learning.

本科二年级
建筑设计（一）：独立居住空间设计
· 刘铨 冷天 吴佳维
课程类型：必修
学时学分：64学时／4学分

Undergraduate Program 2nd Year
ARCHITECTURAL DESIGN 1: INDEPENDENT LIVING SPACE DESIGN · LIU Quan, LENG Tian, WU Jiawei
Type: Required Course
Study Period and Credits: 64 hours/4 credits

课程内容

本次练习的主要任务是综合运用前期案例学习中的知识点——建筑在水平方向上如何利用高度、开洞等操作划分空间；内部空间的功能流线组织及视线关系、墙身、节点、包裹体系、框架结构的构造方式；周围环境对空间、功能、包裹体系的影响等。初步体验一个小型独立居住空间的设计过程。

教学要点

1. 场地与界面：本次设计的场地面积在 80~100 m²，场地单面或相邻两面临街，周边为 1~2 层的传统民居。
2. 功能与空间：本次设计的建筑功能为小型家庭独立式住宅（附设有书房功能）。家庭主要成员包括一对年轻夫妇和 1~2 位儿童（7 岁左右）。新建筑面积 160~200 m²，建筑高度 ≤ 9 m（不设地下空间）。设计者根据设定的家庭成员的职业及兴趣爱好确定空间的功能（职业可以是但不局限于理、工、医、法的技术人员）。
3. 流线组织与出入口设置：考虑建筑内部流线合理性以及建筑出入口与场地周边环境条件的合理衔接。
4. 尺度与感知：建筑中的各功能空间的尺寸需要以人体尺度及人的行为方式作为基本的参照，并通过图示表达空间构成要素与人的空间体验之间的关系。

Course contents

The main task of this exercise is to comprehensively use the knowledge points in the early case study—how to use height, opening and other operations to divide space in the horizontal direction of the building; functional streamline organization and line relationship of internal space; construction modes sight of the wall body, nodes, wrapping system and frame structure; the influence of the surrounding environment on the space, function and wrapping system. And preliminarily experience the design process of a small independent living space.

Teaching essentials

1. Site and interface: The site of this design covers an area of about 80–100 m², facing the street on one side or two adjacent sides, surrounded by 1–2 floors of traditional residential buildings.
2. Function and space: The building function of this design is a small family independent residence (with study function attached). The main members of the family include a young couple and 1–2 children (about 7 years old). The new building area is 160–200 m² and the building height is less than 9 m (no underground space). The designer determines the function of the space according to the set occupations and interests of family members (the occupation can be but not limited to technicians of science, engineering, medicine and law).
3. Streamline organization and entrance and exit setting: Consider the rationality of the internal streamline of the building and the reasonable connection between the entrance and exit of the building and the surrounding environmental conditions of the site.
4. Scale and perception: The size of each functional space in the building needs to take the human body scale and human behaviors as the basic reference, and express the relationship between spatial constituent elements and human spatial experience through diagrams.

本科二年级
建筑设计（二）：校园多功能快递中心设计 · 刘铨 冷天 吴佳维 孟宪川
课程类型：必修
学时学分：64学时／4学分

Undergraduate Program 2nd Year
ARCHITECTURAL DESIGN 2: CAMPUS MULTIFUNCTIONAL EXPRESS CENTER DESIGN · LIU Quan, LENG Tian, WU Jiawei, MENG Xianchuan
Type: Required Course
Study Period and Credits: 64 hours/4 credits

课程内容

在社会信息化、电商化程度日益提高，疫情常态化的背景下，"快递"活跃且丰富地改变了人们的日常生活，甚至某些非常时期，成为保障基本生活需求的重要方式。其中，高校内的快递行为富集，快递服务渐渐成为校园后勤服务中不可或缺的一环，与师生日常活动密不可分，成为校园生活区像食堂、公共浴室、超市一样重要的基础公共设施。本次练习的主要任务，是在南京大学鼓楼校区南园建设一个校园多功能快递服务中心，要求综合运用建筑设计基础课程的知识点，操作一个小型公共建筑设计项目。

Course contents

Against the backdrop of increasing level of social informatization and e-commerce, and the normalization of the epidemic, "express delivery" has actively and richly changed people's daily lives, and even in some extraordinary times, it has become an important way to ensure basic living needs. Wherein, express delivery behaviors in colleges and universities are enriched, and express delivery service has gradually become an indispensable part of campus logistics services. It is inseparable from the daily activities of teachers and students, and has become an important basic public facility in campus living areas like canteens, public bathrooms, and supermarkets. The main task of this exercise is to build a campus multi-functional express service center in the South Park of Gulou Campus of Nanjing University. It is required to comprehensively use the knowledge points of the basic course of architectural design to operate a small public building design project.

本科三年级

建筑设计（三）：专家公寓设计
· 童滋雨　窦平平　黄华青
课程类型：必修
学时学分：72学时 / 4学分

Undergraduate Program 3rd Year
ARCHITECTURAL DESIGN 3: THE EXPERT APARTMENT DESIGN · TONG Ziyu, DOU Pingping, HUANG Huaqing
Type: Required Course
Study Period and Credits: 72 hours/4 credits

课程内容

拟在南京大学鼓楼校区南园宿舍区内新建专家公寓一座，用于国内外专家到访南大开展学术交流活动期间居住。用地位于南园中心喷泉西侧，面积约为3 600 m²。地块上原有建筑将被拆除，新建筑总建筑面积不超过3 000 m²。高度不超过3层。

教学目标

从空间单元到系统的设计训练。

从个体到整体，从单元到体系，是建筑空间组织的一种基本和常用方式。本课题首先关注空间单元的生成，并进一步根据内在的使用逻辑和外在的场地条件，将多个单元通过特定方式与秩序组合起来，形成一个兼备合理性、清晰性和丰富性的整体系统。基本单元的重复、韵律、变异等都是常用的操作手法。

Course contents

It is proposed to build a new expert apartment in the South Park dormitory area of Gulou campus of Nanjing University for domestic and foreign experts to live during their visit to Nanjing University for academic exchange activities. The land is located in the west of the fountain in the center of South Park, covering an area of about 3 600 m². The original buildings on the plot will be demolished, and the total construction area of the new building will not exceed 3 000 m². The height shall not exceed 3 floors.

Teaching objectives

From space units to systematic design training.
From individuals to the whole, from units to the system, is a basic and common way of architectural space organization. This topic first pays attention to the generation of spatial units, and further combines multiple units in order and a specific way according to the internal use logic and external site conditions to form an overall system with rationality, clarity and richness. Repetition, rhythm and variation of basic units are commonly used.

本科三年级

建筑设计（四）：世界文学客厅
· 华晓宁　窦平平　黄华青
课程类型：必修
学时学分：72学时 / 4学分

Undergraduate Program 3rd Year
ARCHITECTURAL DESIGN 4: WORLD LITERATURE LIVING ROOM · HUA Xiaoning, DOU Pingping, HUANG Huaqing
Type: Required Course
Study Period and Credits: 72 hours/4 credits

课程内容

南京古称金陵、白下、建康、建邺……历来是人文荟萃、名家辈出之地，号称"天下文枢"。南京作为六朝古都，亦为中国文学之始。何为文？梁元帝曰："吟咏风谣，流连哀思者，谓之文。"汉魏有文无学，六朝文学《文选》《文心雕龙》《诗品》既是文学评论的开始，也是文学的发端。

2019年，南京入选联合国"世界文学之都"，开展一系列城市空间计划，包括筹建"世界文学客厅"，作为一座以文学为主题的综合性博物馆。该馆选址位于北极阁公园东南隅，用地面积约为5 050 m²，紧临市政府中轴线，毗邻古鸡鸣寺、玄武湖、明城墙、东南大学四牌楼校区等历史文化遗迹，构成城市与山林之间的过渡空间。设计应妥善处理建筑与周边城市环境和既有建筑的关系，彰显中国文学的精神特质。

教学目标

本课程主题是"空间"，学习建筑空间组织的技巧和方法，训练对空间的操作与表达。空间问题是建筑学的基本问题。本课题基于文学主题，训练文本、叙事与空间序列的串联，学生学习空间叙事与空间用途的整体构思，充分考虑人在空间中的行为、空间感受，尝试以空间为手段表达特定的意义和氛围，最终形成一个完整的设计。

Course contents

In 2019, Nanjing was selected as the "capital of world literature" by the United Nations and planned to carry out a series of urban space plans, including preparing to build the "world literature living room" as a comprehensive museum with literature as the theme. The museum is located in the southeast corner of Beijige Park, with a land area of about 5 050 m². It is close to the central axis of the municipal government and adjacent to historical and cultural sites such as ancient Jiming Temple, Xuanwu Lake, Ming City Wall and Sipailou campus of Southeast University, forming a transition space between the city and mountains. The design should properly deal with the relationship between the building and the surrounding urban environment and existing buildings, and highlight the spiritual characteristics of Chinese literature.

Teaching objectives

The theme of this course is "space", to learn the skills and methods of architectural space organization, and train the operation and expression of space. The space problem is the basic problem of architecture. Based on the literary theme, this topic trains the series of the text, narration and spatial sequence, students will learn the overall idea of spatial narration and spatial use, fully consider people's behaviors and spatial feelings in space, try to express specific meaning and atmosphere by means of space, and finally form a complete design.

本科三年级

建筑设计（五）：大学生健身中心改扩建设计

· 傅筱　钟华颖　孟宪川

课程类型：必修

学时学分：64学时 / 4学分

Undergraduate Program 3rd Year

ARCHITECTURAL DESIGN 5: RECONSTRUCTION AND EXPANSION DESIGN OF COLLEGE STUDENT FITNESS CENTER · FU Xiao, ZHONG Huaying, MENG Xianchuan

Type: Required Course

Study Period and Credits: 64 hours/4 credits

教学内容

本学期建筑设计课程训练的主题是"城市建筑"。城市，是建筑与建筑师最重要的舞台。当代复杂多元的城市生活产生了错综复杂的城市物质空间系统，它赋予城市中的建筑诸多限定。而城市中的建筑一旦形成，也即介入和重构了城市物质空间环境，乃至改变和重新定义城市生活本身。在某种程度上，"'事物之间'的形式比事物本身的形式更重要"。这使得建筑设计不再仅仅停留于单纯的自我关注和自我完善，而必须对所在的城市环境做出积极的应对。另一方面，当代城市生活往往要求更大的建筑规模和更为灵活的功能混合，随之而来的是建筑自身的高度复杂性。设计的驱动力同时来自两方面：外在的城市运作系统和内在的建筑运作系统，这两个系统的无缝对接、互动和融合是我们所追求的目标。

Teaching contents

The theme of this semester's architectural design course training is "Urban architecture". Cities are the most important stage for architecture and architects. The contemporary complex and pluralistic urban life has produced a complex system of urban physical space, which gives many limits to the buildings in the city. Once the buildings in the city are formed, they intervene in and reconstructs the physical space environment of the city, and even change and redefine the urban life itself. In a way, "the form 'between things' is more important than the form of things themselves." This makes the architectural design no longer just stay in the simple self-concern and self-improvement, but must make a positive response to the urban environment. Contemporary urban life, on the other hand, often demands a larger architectural scale and a more flexible mix of functions, with this comes a high degree of complexity in the building itself. The driving force of the design comes from two aspects at the same time: the external urban operation system and the internal architectural operation system, and the seamless connection, interaction and integration of the two systems is our goal.

本科三年级

建筑设计（六）：社区文化艺术中心设计

· 张雷　王铠　尹航

课程类型：必修

学时学分：64学时 / 4学分

Undergraduate Program 3nd Year

ARCHITECTURAL DESIGN 6: DESIGN OF COMMUNITY CULTURE AND ART CENTER · ZHANG Lei, WANG Kai, YIN Hang

Type: Required Course

Study Period and Credits: 64 hours/4 credits

课程内容

拟在百子亭风貌区基地处新建社区文化中心，总建筑面积约为 8 000 m²，项目不仅为周边居民提供文化基础设施，同时也期望成为复兴老城的街区活力的文化地标。根据基地条件、功能使用进行建筑和场地设计。总用地详见附图，基地用地面积为 4 600 m²。

设计内容

1. 演艺中心：包含 400 座小剧场，乙级。台口尺寸为 12 m×7 m。根据设计的等级确定前厅、休息厅、观众厅、舞台等面积。观众厅主要为小型话剧及戏剧表演而设置。按 60~80 人化妆布置化妆室及服装道具室，并设 2~4 间小化妆室。要求有合理的舞台及后台布置，应设有排练厅、休息室、候场区以及道具存放间等设施，其余根据需要自定。

2. 文化中心：定位于区级综合性文化站，包括公共图书阅览室、电子阅览室、多功能厅、排练厅以及辅导培训、书画创作等功能室（不少于 8 个且每个功能室面积不低于 30 m²）。

3. 配套商业：包含社区商业以及小型文创主题商业单元。其中社区商业为不小于 200 m² 超市一处，文创主题商业单元面积为 60~200 m²。

4. 其他：变电间、配电间、空调机房、售票、办公、厕所等服务设施根据相关设计规范确定，各个功能区可单独设置，也可统一考虑。地上不考虑机动车停车配建，街区地下统一解决，但需要根据建筑功能面积计算数量。

Course contents

The project plans to build a new community culture and art center at the base of Baiziting historic area, with a total construction area of about 8 000 m². The project not only serves as the cultural infrastructure for surrounding residents, but also hopes to become a cultural landmark to revive the vitality of the old town. The building and site should consider the base conditions and functional use. The total land is shown in the attached figures, and the land area of the base is 4 600 m².

Design contents

1. Performing arts center: It contains a small theatres with 400 seats, class B. The size of the proscenium is 12 m × 7 m. Determine the area of the front hall, lounge, auditorium and stage according to the design level.
2. Cultural center: Serving as the district level comprehensive cultural station, it includes the public reading room, electronic reading room, multi-functional hall, rehearsal hall, counseling and training, calligraphy and painting creation and other functional rooms.
3. Supporting business: It includes community business and small cultural and creative theme business units. Among them, the community business is a supermarket with an area of no less than 200 m², and the area of cultural and creative theme business units is 60-200 m².
4. Others: Service facilities such as the substation room, power distribution room, air conditioning room, ticket, office and toilet are determined according to relevant design specifications.

本科四年级
建筑设计（七）：高层办公楼设计
· 吉国华 王铠 尹航
课程类型：必修
学时学分：64 学时 / 4 学分

Undergraduate Program 4th Year
ARCHITECTURAL DESIGN 7: DESIGN OF HIGH-RISE OFFICE BUILDINGS
· JI Guohua, WANG Kai, YIN Hang
Type: Required Course
Study Period and Credits: 64 hours/4 credits

教学目标

本次课程设计首先希望同学掌握高层建筑的基本特点，研究当代高层建筑的设计策略，了解高层建筑涉及的相关规范与知识，提高综合分析并解决问题的能力。其次还希望同学能主动将建筑与周边文脉、景观等环境要素有机结合，研究环境，在建筑方案中预设某种设计策略，充分发挥创造性，尝试进行某种绿色摩天楼的设计。

教学内容

场地位于在南京市南部新城大校场地块，设计地块在上位规划中确定的性质有办公与商住两种，课程拟减少功能限制，由学生自主调研与策划地块的功能，可建设多种复合功能，具体面积比例自定；合理组织车行、人行入口流线，按不同功能设置出入口集散空间；合理组织场地交通与周边道路关系，合理设置停车场，停车数量（机动车与非机动车）应尽量满足地方法规要求，地下停车场出入口大于 2 个。

Teaching objectives
This course design first hopes that students can master the basic characteristics of high-rise buildings, study the design strategies of contemporary high-rise buildings, understand the relevant norms and knowledge involved in high-rise buildings, and improve the ability of comprehensive analysis and problem solving. Second, it also hopes that students can take the initiative to organically integrate the building with surrounding cultural heritages, landscapes and other environmental elements, study the environment, preset a certain design strategy in the architectural scheme, give full play to creativity, and try to carry out some kind of green skyscraper design.

Teaching contents
The site is located in the Dajiaochang in the south of Nanjing city, and the design plots have two types of properties office buildings and commercial residences determined in the upper planning. The course plans to reduce the functional restrictions, and students will independently investigate and plan the functions of the plots. A variety of composite functions can be built, and the specific area ratio will be determined by themselves. Reasonably organize of vehicle and pedestrian entrance flow lines, according to different functions set up entrance and exit distribution space; The relationship between the site traffic and the surrounding roads should be reasonably organized, and the parking lot should be reasonably set up. The number of parking places should meet the requirements of local laws and regulations as far as possible, and the underground parking lot should have more than 2 entrances and exits.

本科四年级
建筑设计（八）：城市设计
· 童滋雨 胡友培 唐莲
课程类型：必修
学时学分：64 学时 / 4 学分

Undergraduate Program 4th Year
ARCHITECTURAL DESIGN 8: URBAN DESIGN
· TONG Ziyu, HU Youpei, TANG Lian
Type: Required Course
Study Period and Credits: 64 hours/4 credits

教学目标

中国的城市发展已经逐渐从增量扩张转向存量更新。通过对城市建成环境的更新改造而提升环境性能和质量，将成为城市建设的新热点和新常态。与此同时，5G、物联网、无人驾驶等技术的发展又给城市环境的使用方式带来了新的变化。如何在城市更新设计中拓展建筑设计的边界也就成了新的挑战。

城市更新不但需要对建成环境本身有更充分的认知，也要对其中的人流、车流乃至水流、气流等各种动态的活动有正确的认知。从设计上来说，这也大大提高了设计者所面临的问题的复杂性，仅靠个人的直观感受和形式操作难以保证设计的合理性。而借助空间分析、数据统计、算法设计等数字技术，我们可以更好地认知城市形态的特征，理解城市运行的规则，并预测城市未来的发展。通过规则和算法来计算生成城市也是对城市设计思维范式的重要突破。

因此，本次设计将针对这些发展趋势，以城市街巷空间为研究对象，通过思考和推演探索其更新改造的可能性。通过本次设计，学生们可以理解城市设计的相关理论和方法，掌握分析城市形态和创造更好城市环境质量的方法。

Teaching objectives
China's urban development has gradually shifted from incremental expansion to stock renewal. Improving environmental performance and quality through the renewal and transformation of urban built environment will become a new hot spot and new normal of urban construction. At the same time, the development of 5G, Internet of things, unmanned driving and other technologies has brought new changes to the use of urban environment. How to expand the boundary of architectural design in urban renewal design has become a new challenge.
Urban renewal not only needs to have a better understanding of the built environment itself, but also needs a correct understanding of the pedestrian stream, vehicle stream, water stream, air stream and other dynamic activities. In terms of design, it also greatly improves the complexity of problems faced by designers. It is difficult to ensure the rationality of design only by personal intuitive feelings and formal operations. With the help of various digital technologies such as spatial analysis, data statistics and algorithm design, we can better understand the characteristics of the urban form, understand the rules of the urban operation, and predict the future development of the city. Calculating and generating cities through rules and algorithms is also an important breakthrough in the thinking paradigm of urban design.
Therefore, this design will aim at these development trends, take the urban street space as the research object, and explore the possibility of its renewal and transformation through thinking and deduction. Through this design, students can understand the relevant theories and methods of urban design, and master the methods of analyzing the urban form and creating better urban environmental quality.

本科四年级
本科毕业设计
·华晓宁　窦平平　赵潇欣　梁宇舒　施珊珊
鲁安东　史文娟　杨舢　刘铨　童滋雨
王洁琼　王铠　吴蔚　尹航
课程类型：必修
学时学分：1学期/0.75学分

Undergraduate Program 4th Year
GRUDUATION PROJECT
· HUA Xiaoning, DOU Pingping, ZHAO Xiaoxin, LIANG Yushu, SHI Shanshan, LU Andong, SHI Wenjuan, YANG Shan, LIU Quan, TONG Ziyu, WANG Jieqiong, WANG Kai, WU Wei, YIN Hang
Type: Required Course
Study Period and Credits: 1 term /0.75 credit

设计选题
华晓宁．废墟的可能性：国民政府中央广播电台旧址附属建筑及环境改造
窦平平．疗愈环境与"大健康"校园创新设计
赵潇欣．亚太地区"干栏式"建筑研究：以日本列岛为例
梁宇舒，施珊珊．沙漠地区（沙尘暴气候区）乡土建筑室内空气质量研究
鲁安东．高精度空间设计研究
施珊珊．基于建筑室内分区模型的室内空气污染水平及人体暴露分析
史文娟，杨舢，刘铨．居住型历史文化街区—梅园新村：城市更新与社区活化
童滋雨．基于规则和算法的设计和搭建：拓扑互锁结构
王洁琼．源山里景，融文旅情：青岛西海岸杨家山里乡村规划设计
王铠．新城公共基础设施功能更新设计研究
吴蔚．归林·归零：南京晨光1865旧厂房绿色城市农场改造设计
尹航．建筑单体设计＋城市（景观）设计

Design selections
HUA Xiaoning. The possibility of ruin: Annex buildings and environmental renovation of the former site of the National Government Central Radio Station
DOU Pingping. Healing environment and innovative design of "big health" campus
ZHAO Xiaoxin. Study on "stilt house" in asian-pacific region: Taking the japanese archipelago as an example
LIANG Yushu, SHI Shanshan. Research on indoor air quality of vernacular buildings in desert area (sandstorm climate area)
LU Andong. Research on high precision space design
SHI Shanshan. Indoor air pollution level and human exposure analysis based on building indoor partition model
SHI Wenjuan, YANG Jun, LIU Quan. Residential historic and cultural block: Meiyuan New Village: Urban renewal and community activation
TONG Ziyu. Design and construction based on rules and algorithms: Topological interlocking structure
WANG Jieqiong. The source of the scenery in the mountains, the integration of culture and travel: The planning and design of Yangjiashan Village on the west coast of Qingdao
WANG Kai. Research on functional renewal design of new town public infrastructure
WU Wei. Returning to the forest · returning to zero: Nanjing Chenguang 1865 Old Factory Building green urban farm renovation design
YIN Hang. Single building design + urban (landscape) design

研究生一年级
建筑设计研究（一）：基本设计
·周凌
课程类型：必修
学时学分：40学时/2学分

Graduate Program 1st Year
ARCHITECTURAL DESIGN RESEARCH 1: BASIC DESIGN · ZHOU Ling
Type: Required Course
Study Period and Credits: 40 hours/2 credits

研究课题
西连岛北片区更新改造规划及建筑设计

课程简介
连岛位于江苏、山东两省沿海交界处海域，行政建制隶属连云港市连云区，与连云港城区隔海相望。连岛东西长约5.5 km，南北最宽处约1.8 km。海岸线长约18 km，礁石岸约8 km。连岛地形陡峭，起伏较大，山形错落有致。岛上植物种类丰富，植被密布，林地面积超过4.5万 km²，森林覆盖率达65%。
本次设计分为片区更新改造规划与单体建筑设计两个部分，旨在培养学生依托生态、高效的片区更新理念，对单体建筑进行深入设计的能力。同时学习生态规划、业态策划、产业布局等知识。

设计内容
3~4人一组进行设计：
任务一：片区更新改造规划。
任务二：单体建筑设计。
（四选一：城市客厅、海滨精品酒店集群、半岛星级酒店、山上民居组团）

Research subject
Renovation planning and architectural design of the northern area of West Liandao

Course descriptions
Liandao is located in the sea area at the junction of the coastal areas of Jiangsu and Shandong provinces. Its administrative system is subordinate to Lianyun District of Lianyungang City, and it faces Lianyungang City across the sea. Liandao is about 5.5 kilometers long from east to west, and about 1.8 kilometers wide at its widest place from north to south. The coastline is about 18 kilometers long, and the reef shoreline is about 8 kilometers. The terrain of Liandao is steep, with large ups and downs, and the mountain forms are well arranged and proportioned. The island is rich in plant species and densely covered with vegetation. The forest area exceeds 45 000 square kilometers, and the forest coverage rate reaches 65%.
This design is divided into two parts: area renewal planning and individual building design, aiming at cultivating students' ability to carry out in-depth design of individual buildings relying on the concept of ecological and efficient area renewal. At the same time, they can learn ecological planning, business planning, industrial layout and other knowledge.

Design contents
Design in groups of 3-4:
Task 1: Area renewal planning.
Task 2: Individual building design.
(Choose one from four: city living room, seaside boutique hotel cluster, peninsula star hotel, mountain dwelling group)

研究生一年级
建筑设计研究（一）：基本设计
· 傅筱
课程类型：必修
学时学分：40学时 / 2学分

Graduate Program 1st Year
ARCHITECTURAL DESIGN RESEARCH 1: BASIC DESIGN
· FU Xiao
Type: Required Course
Study Period and Credits: 40 hours/2 credits

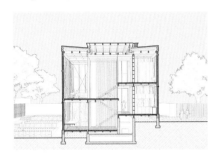

课程内容
建筑要求
1. 建筑基底面积不得超过130 m²。要求内部布局紧凑经济，使用功能合理，在满足功能需求之下，尽量减少面积以节省造价，总建筑面积不宜超过220 m²。
2. 可选用结构为：钢筋混凝土框架结构(或剪力墙结构)、砖混结构、轻钢龙骨结构体系（或轻钢体系）、轻木框架结构体系（或轻木龙骨体系），同一地块的小组不得选用相同的结构体系。外墙材料选择需与结构体系有一定的关联性，并需要考虑保温隔热要求。
3. 主要入户空间要求朝南或者朝东。需配备服务入口，服务入口朝向自定。
4. 空调形式为小型一拖多VRV空调，需设计放置位置，并考虑室内出风口位置。
5. 明厨明卫，需进行烟囱设计。
6. 具体房间数量要求，面积自定。
（1）客厅；（2）餐厅；（3）厨房；（4）主卧1间（带卫生间，考虑小孩同居，步入式衣橱）；（5）客卧1间（使用公用卫生间）；（6）客卧1间（使用公用卫生间，未来作为子卧）；（7）丈夫工作空间；（8）公用卫生间，根据需要确定数量；（9）洗衣空间；（10）储藏空间。
技术要求：鼓励用BIM（Revit）、Enscape设计和表达。
参考书目：《加拿大木框架房屋建筑》（学院资料室）、各类住宅设计书籍。

Course contents
building requirements
1. The building base area shall not exceed 130 m². It is required that the internal layout is compact and economical, and the use functions are reasonable. Under the condition of meeting the functional requirements, the area should be reduced as much as possible to save the cost. The total construction area should not exceed 220 m².
2. The optional structures are: reinforced concrete frame structure (or shear wall structure), brick-concrete structure, light steel keel structure system (or light steel system), light wood frame structure system (or light wood keel system). Groups of the same block must not use the same structural system. The selection of exterior wall materials needs to have a certain relationship with the structural system, and the thermal insulation requirements need to be considered.
3. The main entrance space is required to face south or east. It needs to be equipped with a service entrance, and the orientation of the service entrance can be determined by students.
4. The air conditioner is a small multi split VRV air conditioner, and the placement location needs to be designed, taking into account the location of the indoor air outlet.
5. The kitchen and bathroom should have windows, chimney design is required.
6. The specific number of rooms, and the area are self-determined.

研究生一年级
建筑设计研究（一）：基本设计
· 冷天
课程类型：必修
学时学分：40学时 / 2学分

Graduate Program 1st Year
ARCHITECTURAL DESIGN RESEARCH 1: BASIC DESIGN · LENG Tian
Type: Required Course
Study Period and Credits: 40 hours/2 credits

课程内容
校园场所微改造

作为遗产保护领域中的永恒矛盾，保护与利用的互动关系深刻地影响着建筑师对待历史建筑实践的态度。当下中国的城市建设发展，业已从"增量开发"走向"存量更新"。其中，大量历史性建筑（文物建筑、历史建筑、文化遗产等），都面临着如何在保存特有历史文化价值的前提下，充分活化利用其原有空间（内部、外部）的难题。本课程以南京大学鼓楼校区中的历史建筑、场所为操作对象，引导学生理解设计的逻辑性与综合性，综合考虑校园历史文脉、现状格局和未来发展等多方面因素，从创造性的概念构思和建筑学的基本问题出发，对校园中规模较小的建筑、场所等进行深入的改造设计，最终在历史和现实之间取得平衡，通过一个富有创造性的设计来激发既有空间的活力。

Course contents
Micro-renovation of campus places
As an eternal contradiction in the field of heritage protection, the interactive relationship between protection and utilization has profoundly affected architects' attitudes towards historical architectural practices. The current urban construction and development in China has shifted from "incremental development" to "stock renewal". A large number of historic buildings (cultural relics, historical buildings, cultural heritage, etc.) are faced with the problem of how to fully activate and utilize the original spaces (inside and outside) while preserving their unique historical and cultural values. This course takes the historical buildings and places in the Gulou Campus of Nanjing University as the operating objects, guides students to understand the logic and comprehensiveness of the design, comprehensively considers various factors such as the campus's historical context, current situation and future development, and conceives from a creative concept. Starting from the basic issues of architecture, we will carry out in-depth renovation and design of small-scale buildings and places in the campus, and finally achieve a balance between history and reality, and stimulate the vitality of the existing space through a creative design.

研究生一年级
建筑设计研究（一）：概念设计
· 鲁安东
课程类型：必修
学时学分：40 学时 / 2 学分

Graduate Program 1st Year
ARCHITECTURAL DESIGN RESEARCH 1: CONCEPTUAL DESIGN · LU Andong
Type: Required Course
Study Period and Credits: 40 hours/2 credits

课程内容

研究对象：作为记忆介质的建筑学

长久以来，建筑在城市中承担着集体记忆的介质功能。而在当代，数字技术在日常生活中的普遍运用，从根本上改变了城市记忆的塑造形式和接受形式。建筑作为城市记忆的枢纽介质，必须整合当代城市记忆的技术架构。随着"记忆的艺术"逐渐被更替为"记忆的技术"，这也带来了全新的创造可能，同时也要求对"设计"这一工作进行重新定义。此次课程将从城市记忆的空间介质入手，对城市轴线、物体性、景象／图像、叙事、空间仪轨等中介机制进行研究，进而提出一个针对性的设计计划。

课程要求

学生人数：18 人，三人一组。
作业要求：各组自定研究选题，并针对选题拟定任务书，完成相关技术图纸绘制。
每组完成概念手册一份（80~100 页）。

Course contents

Research object: Architecture as a medium of memory
For a long time, architecture has assumed the media function of collective memory in the city. In contemporary times, the widespread use of digital technology in daily life has fundamentally changed the molding form and acceptance form of urban memory. As the pivotal medium of urban memory, architecture must integrate the technical framework of contemporary urban memory. As the "art of memory" is gradually replaced by "technology of memory", it also brings new possibilities for creation, and at the same time requires a redefinition of the work of "design". This course will start with the spatial medium of urban memory, conduct research on intermediary mechanisms such as the urban axis, objectivity, scene/image, narrative, and spatial rituals, and then propose a targeted design plan.

Course requirements

Number of students: 18, each group of three.
Homework requirements: Each group decides the research topic by students, draws up a briek for the selected topic, and completes the drawing of relevant technical drawings. Each group completes a concept booklet (80–100 pages).

研究生一年级
建筑设计研究（一）：概念设计
· 周渐佳
课程类型：必修
学时学分：40 学时 / 2 学分

Graduate Program 1st Year
ARCHITECTURAL DESIGN RESEARCH 1: CONCEPTUAL DESIGN
· ZHOU Jianjia
Type: Required Course
Study Period and Credits: 40 hours/2 credits

课程内容

2020 年的硕士生概念设计受线上空间的启发，以"从物理空间到线上空间"为主题提取空间原型加以设计。线上空间曾经被认为是物理空间的附属，却在很短的时间内成为所有活动发生的重要乃至唯一载体。这个过程也将线上空间推到了建筑学科的面前，我们会发现线上空间是与物理空间并行，且有着同等意义的一个新领域，此前却极少获得来建筑学科的关注。2021 年"元宇宙"概念对所有行业形成了冲击，也使得更多可能性展现在我们面前。顺延这条脉络的讨论正当其时。本学期的硕士生概念设计以"从线下到线上"为主题，形成一个开放的讨论过程。课程以建筑系学最熟悉的"studio"为基本的空间类型，在八周的设计、理论教学中围绕与之相关的线下、线上的概念、行动等展开探究。一方面提供相应的技术支持，另一方面通过系列讲座进一步打开思考的广度与深度。最终成果同样形成展览，在线上与线下空间同时展示。课程中的所有讲座、小论文、讨论、手稿将汇编成册，作为对课程的记录。

Course contents

Inspired by the online space, the conceptual design of the master students in 2020 is based on the theme of "from physical space to online space" to extract the space prototype and design. The online space was once considered an adjunct to the physical space, but it has become an important and even the only carrier for all activities in a short period of time. This process also pushes the online space to the front of the architectural discipline. We will find that the online space is a new field that is parallel to the physical space and has the same meaning. It has received little attention from the architectural discipline before. In 2021, the concept of "Metaverse" had an impact on all industries, and more possibilities will be presented to us. It is timely to discuss on the extension of this thread. The concept design of master students in this semester takes "from offline to online" as the theme, forming an open discussion process. The course uses the most familiar "studio" as the basic space type in the Department of Architecture, and explores related offline and online concepts and actions during the eight-week design and theoretical teaching. On the one hand, it provides corresponding technical support, on the other hand, it further opens up the breadth and depth of thinking through a series of lectures. The final results also form an exhibition, which is displayed both online and offline. All lectures, essays, discussions, and manuscripts in the course will be compiled into a book as a record of the course.

研究生一年级
建筑设计研究（一）：城市设计
・华晓宁
课程类型：必修
学时学分：40 学时 / 2 学分

Graduate Program 1st Year
ARCHITECTURAL DESIGN RESEARCH 2: URBAN DESIGN · HUA Xiaoning
Type: Required Course
Study Period and Credits: 40 hours/2 credits

课程议题
基础设施是当代城市研究与实践的重要主题。作为为社会生产和居民生活提供公共服务的物质工程设施及其系统，它保障着城市有机体的运行，同时又是城市物质空间系统的重要组成部分，自身便占据了场址，界定了空间，形成了场所，连接成系统，构筑了场域。

被传统建筑学忽视多年，许多基础设施成为城市中消极和被动的要素。随着城市的演进与变迁，既往的基础设施场址面临着再生。然而，在漫长的既往，基础设施早已沉浸在多重的城市文脉中。如何在复杂的城市多重文脉中重定义基础设施，将其视为重要的城市操作性对象与媒介，对基础设施"赋能"并转化为城市中更为积极、能动的场所，激发城市活力，是本课题的主要目标。

课程要求
对中华门火车站及其周边地段（东至雨花路，西至虹悦城，南到雨花西路、雨花东路，北至应天高架）进行深入调研，分析存在问题与矛盾，构想该区段未来愿景，自行拟定任务书，提出改造更新策略，完成方案设计。方案需将城市、建筑、环境景观综合考虑，进行整合设计。

Course topic
Infrastructure is an important theme in contemporary urban research and practice. As a material engineering facility and its system to provide public services for social production and residents' life, it guarantees the operation of the urban organism, and at the same time is an important part of the urban material space system, which occupies the site, defines the space, forms the place, connects into a system, and constructs the field.

Neglected by traditional architecture for many years, much infrastructure becomes a passive and passive element in the city. As cities evolve and change, former infrastructure sites face regeneration. However, over the long past, infrastructure has been immersed in multiple urban contexts. How to redefine infrastructure in complex urban contexts, treat it as an important urban operable medium (operable medium), and "enable" infrastructure and transform it into a more active and dynamic place in the city, stimulating urban vitality, is the main goal of this course.

Course requirements
Conduct in-depth investigation on Zhonghuamen Railway Station and its surrounding areas (east to Yuhua Road, west to Hongyue City, south to Yuhua West Road, Yuhua East Road, north to Yingtian Elevated Road), analyze the existing problems and contradictions, conceive the future vision of the section, formulate the assignment, propose the transformation and renewal strategy, and complete the scheme design. The scheme needs to comprehensively consider the urban, architectural and environmental landscapes and carry out integrated design.

研究生一年级
建筑设计研究（一）：城市设计
・胡友培
课程类型：必修
学时学分：40 学时 / 2 学分

Graduate Program 1st Year
ARCHITECTURAL DESIGN RESEARCH 2: URBAN DESIGN · HU Youpei
Type: Required Course
Study Period and Credits: 40 hours/2 credits

问题与任务
长三角地区是中国城市化程度最高的区域之一，构成了巨大的城市群落与延绵的都市区。在这里，格网城市化模式以巨大的吞吐量吞噬着千百年逐渐形成的人居与自然景观，不断占领多样与差异化的地表，塑造出一座座类似的新城、居住点。都市区中的生活被标准化在格网的城市中，单调而空洞。

很大程度上，我们仍然沿用着现代主义城市规划的草纸策略，无差别地对待着都市区复杂的自然与村庄系统。在可持续、生态理念已成为普遍共识的今天，这个草纸的本底似乎并未有多大改观。尽管格网的城市化是一种高效、低成本的空间生产技术，但其弊端也同样显著并且显得陈词滥调。可能在格网模型化之外，在中心城区外围广袤的都市区中，能够想象另一种城市化模型及其建筑学？

Issues and tasks
The Yangtze River Delta region is one of the most urbanized regions in China, constituting a huge urban agglomeration and a stretch of urban areas. Here, the grid urbanization model devours the human settlements and natural landscapes with huge throughput that have been gradually formed over thousands of years, constantly occupying diverse and differentiated surfaces, and shaping similar new cities and settlements. Life in metropolitan areas is standardized in a grid of cities, monotonous and empty.

To a large extent, we still follow the sketch strategy of modernist urban planning, which treats the complex natural and rural systems of urban areas without discrimination. Today, when the concept of sustainability and ecology has become a general consensus, the background of this sketch does not seem to have changed much. Although the urbanization of grids is an efficient, low-cost space production technology, its drawbacks are just as significant and cliched. Could it be that in addition to grid modeling, in a vast metropolitan area outside the central urban area, imagine another model of urbanization and its architecture?

研究生一年级
建筑设计研究（二）：综合设计
· 程向阳
课程类型：必修
学时学分：40 学时 / 2 学分

Graduate Program 1st Year
ARCHITECTURAL DESIGN RESEARCH 2: COMPREHENSIVE DESIGN
· CHENG Xiangyang
Type: Required Course
Study Period and Credits: 40 hours/2 credits

课程内容

中国科学院南京土壤研究所位于南京市玄武区北京东路71号，成立于1953年，其前身为1930年创立的中央地质调查所土壤研究室。中国科学院南京地理与湖泊研究所位于南京市玄武区北京东路73号，其前身系1940年8月成立的中国地理研究所，1988年1月改名为中国科学院南京地理与湖泊研究所并沿用至今，是中国唯一以湖泊—流域系统为主要研究对象的中国地理研究所。上述两个研究所共用一个大院空间，没有明确的物理边界分割，总占地面积 34 760 m²。

设计任务

在相关调研、分析和上位规划基本条件前提下，在场地范围内规划设计一个以营造科研、教育、文化或社区配套等为主体内容的开放城市公共空间场所。由设计者研究自定主体内容以及其它兼容业态，如休闲娱乐、酒店、商业、商务以及公寓等。

规划设计除两个历史建筑必须保留外，应最大化利用现有建筑空间价值，尊重现有重要的空间脉络，设计者可结合规划设计，自定未来改造用途以及改造力度。

Course contents

The Institute of Soil Science, Chinese Academy of Sciences is located at No. 71, Beijing East Road, Xuanwu District, Nanjing. The Soil Research Institute was established in 1953 which was formerly known as the Soil Research Office of the Central Geological Survey Institute found in 1930. The Nanjing Institute of Geography and Limnology of the Chinese Academy of Sciences is located at No. 73, Beijing East Road, Xuanwu District, Nanjing which was formerly known as the Chinese Institute of Geography, found in 1940, August. It was renamed the Institute of Earth and Lakes in 1988, January and is still in use today. It is the only Chinese institute of Geography that focuses on lake-watershed systems as its main research object. The above-mentioned two research institutes share a compound space without clear physical boundary division, covering a total area of 34 760 m².

Design tasks

Under the premise of relevant research, analysis and upper-level planning, plan and design an open urban public space within the scope of the site, with the main content of scientific research, education, culture or community facilities. The designer studies and defines main content and other compatible formats, such as leisure and entertainment, hotels, commerce, business and apartments, etc..
In addition to the two historical buildings that must be preserved, the planning and design should maximize the use of the existing architectural space value and respect the existing important space. The designer can combine planning and design to determine the future renovation use and renovation intensity.

研究生一年级
建筑设计研究（二）：综合设计
· 金鑫
课程类型：必修
学时学分：40 学时 / 2 学分

Graduate Program 1st Year
ARCHITECTURAL DESIGN RESEARCH 1: COMPREHENSIVE DESIGN · JIN Xin
Type: Required Course
Study Period and Credits: 40 hours/2 credits

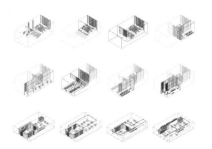

课程内容

1. 尺度转换

工业建筑、构筑物和场地等通常具有远超人体尺度的巨大体量，并容纳大量复杂的机器和设备的运作。这样的物质环境，具有机器化、非人性的尺度和空间，难以与常人的生活、工作等活动相关联。以空间的再生作为设计研究的核心，意味着将工业建筑的巨大空间转向民用、公共的空间。

而工业建筑巨大的空间尺度和坚固的结构体系，提供了重新组织交通流线的可能。这同样需要作相应的空间尺度转换研究，以汽车和车行的尺度作为基本单元来研究工业建筑的空间适应性。

2. 程序重置

在原为满足生产工艺流程要求而设置的工业建筑空间及其组织关系中，重新置入符合城市生活需求的新的程序与功能，合理安排新的活动内容。

3. 结构重组

在工业建筑的改造中，为了满足空间再利用的需求，可对工业建筑既有结构体系进行改变或重组。新置入的结构体与既有工业建筑结构体系可能形成多种空间位置和受力关系。

Course contents

1. Scale conversion

Industrial buildings, structures, and sites usually have huge volumes that far exceed the scale of the human body, and accommodate the operation of a large number of complex machines and equipment. Such a physical environment has a mechanized and inhuman scale and space, and is difficult to relate to ordinary people's life, work and other activities. Taking the regeneration of space as the core of design research means turning the huge space of industrial buildings into civil and public spaces.
The huge spatial scale and solid structural system of industrial buildings provide the possibility to reorganize the traffic flow. This also requires corresponding research on spatial scale conversion, using the scale of cars and car dealerships as the basic unit to study the spatial adaptability of industrial buildings.
2. Program reset
In the industrial building space and its organizational relationship originally set up to meet the requirements of the production process, new procedures and functions that meet the needs of urban life are re-installed, and new activities are reasonably arranged.
3. Restructuring
In the transformation of industrial buildings, in order to meet the needs of space reuse, the existing structural system of industrial buildings can be changed or reorganized. The newly placed structure may form a variety of spatial positions and stress relationships with the existing industrial building structure system.

研究生国际教学工作坊
光几何
· 约瑟夫·施瓦茨
课程类型：选修
学时学分：18 学时 / 1 学分

Postgraduate International Design Studio
Light Geometries · Joseph SCHWARTZ
Type: Elective Course
Study Period and Credits: 18 hours/1 credit

背景介绍
该工作坊提出了一种新颖的设计策略，旨在将建筑、结构和太阳能控制方面整合为概念设计阶段的组成部分。因此，建筑师和工程师可以在设计过程的早期做出明智的设计决策，以平衡设计的技术和创意方面。基于几何图形的方法，仅考虑几何设计参数，是所提出的整体设计方法的核心，因为它们允许设计人员直接理解单个设计参数对结构和太阳能性能的影响，而不会忽视建筑考虑因素。

教学目标
该工作坊旨在向建筑专业的学生介绍基于几何的结构和太阳能设计工具，并将他们的设计进一步发展为展馆。学生将使用基于几何图形的方法及其数字化实现来调查建筑、结构和太阳能控制方面。在工作坊使用图形静力学和应力场来设计整体结构行为以及定制的互锁木材到木材连接。此外，用于太阳能控制的图形方法将为设计有效的遮阳设备提供信息。最后，人类行为被纳入这种基于几何形状的设计过程，将结构与特定的人类活动联系起来，从而定义展馆的功能和空间标准。

Background introduction
This studio proposes a novel design strategy aimed at integrating architectural, structural, and solar control aspects as integral parts of the conceptual design phase. As such, architects and engineers can make informed design decisions early in the design process in an attempt to balance the technical and creative sides of design. Geometry-based graphical methods, which consider only geometric design parameters, are the core of the proposed holistic design approach as they allow the designer to directly comprehend the effect of a single design parameter on structural and solar performance without neglecting architectural considerations.

Teaching objectives
The studio aims to introduce architecture students to geometry-based tools for structural and solar design, and further develop their design into a pavilion. In the studio, students will examine architectural, structural, and solar control aspects using geometry-based graphical methods and their digital implementations. Graphic statics and stress fields are used in the studio to design the global structural behaviour as well as bespoke interlocking timber-to-timber connections. Moreover, graphical methods for solar control will inform the design of effective sun-shading devices. Finally, human behaviour is incorporated into this geometry-based design process to connect structures to specific human activities, thereby defining the pavilion's functional and spatial criteria.

研究生国际教学工作坊
可持续的城市密度
· 索布鲁赫·赫顿 王洁琼
课程类型：选修
学时学分：18 学时 / 1 学分

Postgraduate International Design Studio
Sustainable Urban Densification
· Sauerbruch HUTTON, WANG Jieqiong
Type: Elective Course
Study Period and Credits: 18 hours/1 credit

教学内容
任务将集中于南京民用无线电短波大楼的开发，见附件。

要求构思场地的可持续城市发展，包括现有建筑并使场地致密，同时保持公园，例如它的绿色品质，尽可能与周围环境融为一体。

学生将被要求定义一个程序，并为场地的致密化提出合理的建议，同时结合现有建筑的修复／改建和新建筑的概念。可持续性将成为设计指南，融入城市战略和建筑设计。学生还将被要求考虑使用可回收材料和施工方法。

本次工作坊将支持学生在中国城市／社会背景下探索不同维度和层次的可持续发展的各种表达方式。

第一个主题应该是对"绿色城市"的批判性观点，然后要求学生批判性地研究建筑物的生命周期。首先，他们将要求审查建筑材料，特别是木材材料的使用。其次，他们需要考虑施工方法。一种方法可能是模块化系统的高级使用，模块化系统在架构方面具有悠久的历史，并且在灵活性、可回收性和生产方面可能是有益的。第三，邀请他们思考建筑的第二次生命——未来的用途变化。

Teaching contents
The task will focus on developing the Civil Radio Shortwave Building in Nanjing, see attachment.
The task is to conceive a sustainable urban development of the site that includes the existing buildings and densities the site while maintaining the park i.e. green qualities of it, integrating as good as possible in the surroundings.
Students will be requested to define a program and make a reasonable proposal for the densification of the site while combining restoration/conversion of the existing buildings and the conception of new ones. Sustainability would be the design guide, integrated in the urban strategy and architecture design. Students will also be asked to consider the use of recyclable materials and methods of construction.
The studio will support students to explore various expression of sustainability in different dimensions and levels in China urban/society context.
First topic should be the critical view on a "green city". Then the students are requested to critically look into the life cycle of a building. First they would be asked to review construction materials, especially the use of timber. Second, they need to think about the construction method. One method could be the advanced use of modular systems, which have quite a history in structure and could be beneficial in terms of flexibility, recyclability and production. Third, they are invited to think about the second life of a building—the future change of use.

建筑理论课程
ARCHITECTURAL THEORY COURSES

本科二年级
建筑导论 · 赵辰 等
课程类型：必修
学时 / 学分：36 学时 / 2 学分

Undergraduate Program 2nd Year
INTRODUCTOTY GUIDE TO ARCHITECTURE
• ZHAO Chen, et al.
Type: Required Course
Study Period and Credits: 36 hours / 2 credits

课程内容
1. 建筑学的基本定义
第一讲：建筑与设计 / 赵辰
第二讲：建筑与城市 / 丁沃沃
第三讲：建筑与生活 / 张雷
2. 建筑的基本构成
（1）建筑的物质构成
第四讲：建筑的物质环境 / 赵辰
第五讲：建筑与节能技术 / 郜志
第六讲：建筑与生态环境 / 吴蔚
第七讲：建筑的环境智慧 / 窦平平
（2）建筑的文化构成
第八讲：建筑与人文、艺术、审美 / 赵辰
第九讲：建筑与环境景观 / 华晓宁
第十讲：中西方风景观念与设计 / 史文娟
第十一讲：建筑与身体经验 / 鲁安东
（3）建筑师职业与建筑学术
第十二讲：建筑与表现 / 赵辰
第十三讲：建筑与乡村复兴 / 黄华青
第十四讲：建筑与数字技术 / 钟华颖
第十五讲：建筑师的职业技能与社会责任 / 傅筱

Course contents
1. Basic definition of architecture
Lecture 1: Architecture and design / ZHAO Chen
Lecture 2: Architecture and urbanization / DING Wowo
Lecture 3: Architecture and life / ZHANG Lei
2. Basic attributes of architecture
(1) Physical attributes
Lecture 4: Physical environment of architecture / ZHAO Chen
Lecture 5: Architecture and energy saving / GAO Zhi
Lecture 6: Architecture and ecological environment / WU Wei
Lecture 7: Environmental intelligence in architecture / DOU Pingping
(2) Cultural attributes
Lecture 8: Architecture and civilization, arts, aesthetic / ZHAO Chen
Lecture 9: Architecture and landscaping environment / HUA Xiaoning
Lecture 10: Landscaping View and Design in Comparison of China and West / SHI Wenjuan
Lecture 11: Architecture and body experience / LU Andong
(3) Architects: Profession and academy
Lecture 12: Architecture and presentation / ZHAO Chen
Lecture 13: Architecture and rural rival / HUANG Huaqing
Lecture 14: Architectural and digital technology / ZHONG Huaying
Lecture 15: Architect's professional technique and social responsibility / FU Xiao

---

本科二年级
建筑设计基本原理 · 周凌
课程类型：必修
学时 / 学分：36 学时 / 2 学分

Undergraduate Program 2nd Year
BASIC THEORY OF ARCHITECTURAL DESIGN • ZHOU Ling
Type: Required Course
Study Period and Credits: 36 hours / 2 credits

教学目标
本课程是建筑学专业本科生的专业基础理论课程。本课程的任务主要是介绍建筑设计中形式与类型的基本原理。形式原理包含历史上各个时期的设计原则，类型原理讨论不同类型建筑的设计原理。

课程要求
1. 讲授大纲的重点内容；
2. 通过分析实例启迪学生的思维，加深学生对有关理论及其应用、工程实例等内容的理解；
3. 通过对实例的讨论，引导学生运用所学的专业理论知识，分析、解决实际问题。

课程内容
1. 形式与类型概述
2. 古典建筑形式语言
3. 现代建筑形式语言
4. 当代建筑形式语言
5. 类型设计
6. 材料与建造
7. 技术与规范
8. 课程总结

Teaching objectives
This course is a professional basic theory course for the undergraduate students of architecture. The main task of this course is to introduce the basic principles of the form and type in architectural design. Form theory contains design principles in various periods of architecture history, type theory discusses the design principles of different types of buildings.

Course requirements
1. Teach the key elements of the outline;
2. Enlighten students' thinking and enhance students' understanding of the theories, applications and project examples through analyzing examples;
3. Help students to use the professional knowledge to analyse and solve practical problems through the discussion of examples.

Course contents
1. Overview of forms and types
2. Classical architecture form language
3. Modern architecture form language
4. Contemporary architecture form language
5. Type design
6. Materials and construction
7. Technology and specifications
8. Course summary

本科三年级
居住建筑设计与居住区规划原理 · 冷天 刘铨
课程类型：必修
学时 / 学分：36 学时 /2 学分

Undergraduate Program 3rd Year
THEORY OF HOUSING DESIGN AND RESIDENTIAL PLANNING · LENG Tian, LIU Quan
Type: Required Course
Study Period and Credits: 36 hours / 2 credits

课程内容
第一讲：课程概述
第二讲：居住建筑的演变
第三讲：套型空间的设计
第四讲：套型空间的组合与单体设计（一）
第五讲：套型空间的组合与单体设计（二）
第六讲：居住建筑的结构、设备与施工
第七讲：专题讲座：住宅的适应性，支撑体住宅
第八讲：城市规划理论概述
第九讲：现代居住区规划的发展历程
第十讲：居住区的空间组织
第十一讲：居住区的道路交通系统规划与设计
第十二讲：居住区的绿地景观系统规划与设计
第十三讲：居住区公共设施规划、竖向设计与管线综合
第十四讲：专题讲座：房地产开发
第十五讲：专题讲座：住区规划和房屋设计实践
第十六讲：课程总结，考试答疑

Course contents
Lecture 1: Introduction of the course
Lecture 2: Development of residential buildings
Lecture 3: Design of dwelling space
Lecture 4: Dwelling space arrangement and monomer building design (1)
Lecture 5: Dwelling space arrangement and monomer building design (2)
Lecture 6: Structure, facilities and construction of residential buildings
Lecture 7: Special lecture: Adaptability of residential buildings, supporting houses
Lecture 8: Introduction of theories of urban planning
Lecture 9: History of modern residential area planning
Lecture 10: Organization of residential space
Lecture 11: Traffic system planning and design of residential areas
Lecture 12: Landscape system planning and design of residential areas
Lecture 13: Public facilities and infrastructure system
Lecture 14: Special lecture: Real estate development
Lecture 15: Special lecture: The practice of residential planning and housing design
Lecture 16: Summary, question of the test

本科四年级
建筑设计行业知识与创新实践 · 周凌 梁宇舒
课程类型：选修
学时 / 学分：36 学时 /2 学分

Undergraduate Program 4st Year
Knowledge and Innovative Practice in the Architectural Design Industry · ZHOU Ling, LIAGN Yushu
Type: Elective Course
Study Period and Credits: 36 hours / 2 credits

课程内容
建筑设计是一项应用性很强，科学、艺术、技术、人文综合性很强的学科与工作。本课程以兼顾本科毕业生及研究生的本硕贯通课、同时是面向校内其他专业同学的公共选修课为定位，作为南京大学校级"创新创业"立项课程，教学注重学生学习方法、创新能力、主动性思考、沟通表达能力和素质的培养，将建筑学背景下的行业知识与创新实践相结合，为学生开拓行业认知，了解前沿的行业知识，提供多元的行业发展视野，旨在建设面向院外的高质量开放课程。课程同步面向公众进行线上推广，逐步建设南京大学建筑与城市规划学院大师课程库。

Course contents
Architectural design is a highly applied discipline and work with strong integration of science, art, technology and humanities. This course is designed to accommodate both undergraduate and graduate students in a master's program, as well as a public elective course for other majors on campus and a project of Nanjing University. It focuses on the cultivation of students' learning method. innovation ability, initiative thinking, communication and expression ability and quality. It combines industry knowledge and innovative practice in the context of architecture, develops industry awareness for students, and provides them with cutting-edge industry knowledge and a diversified vision of the development of the industry, with the aim of building a high-quality open course for people whole don't study in the university. The course will be synchronized for online promotion to the public, and the master courses library of the School of Architecture and Urban Planning of Nanjing University is gradually constructed.

研究生一年级
现代建筑设计基础理论 · 周凌，梁宇舒
课程类型：必修
学时 / 学分：18 学时 /1 学分

Graduate Program 1st Year
PRELIMINARIES IN MODERN ARCHITECTURAL DESIGN • ZHOU Ling, LIANG Yushu
Type: Required Course
Study Period and Credits: 18 hours/1 credit

教学目标
　　建筑可以被抽象到最基本的空间围合状态来面对它所必须解决的基本的适用问题，用最合理、最直接的空间组织和建造方式去解决问题，以普通材料和通用方法去回应复杂的使用要求，是建筑设计所应该关注的基本原则。

课程要求
1. 讲授大纲的重点内容；
2. 通过分析实例启迪学生的思维，加深学生对有关理论及其应用、工程实例等内容的理解；
3. 通过对实例的讨论，引导学生运用所学的专业理论知识，分析、解决实际问题。

课程内容
1. 基本建筑的思想
2. 基本空间的组织
3. 建筑类型的抽象与还原
4. 材料的运用与建造问题
5. 场所的形成及其意义
6. 建筑构思与设计概念

Teaching objectives
Architecture can be abstracted into the spatial enclosure state to encounter basic application problems which must be settled. Solving problems with most reasonable and direct spatial organization and construction mode, and responding to operating requirements with common materials and general methods are basic principle concerned by building design.

Course requirements
1. To teach key contents of the syllabus;
2. To inspire students' thinking, deepen students' understanding on contents such as relevant theories and applications and engineering examples through case analysis.
3. To help students to use professional theories to analyze and solve practical problems through discussion of instances.

Course contents
1. Basic architectural thought
2. Basic spatial organization
3. Abstraction and restoration of architectural types
4. Utilization and construction of materials
5. Formation of the site and its meaning
6. Architectural conception and design concepts

研究生一年级
建筑与规划研究方法 · 罗小龙　鲁安东　等
课程类型：必修
学时 / 学分：18 学时 /1 学分

Graduate Program 1st Year
RESEARCH METHOD OF ARCHITECTURE AND URBAN PLANING
• LUO Xiaolong, LU Andong, et al.
Type: Required Course
Study Period and Credits: 18 hours/1 credit

教学目标
　　面向学术型硕士的必修课程。它将向学生全面地介绍学术研究的特性、思维方式、常见方法以及开展学术研究必要的工作方式和写作规范。考虑到不同领域研究方法的差异，本课程的授课和作业将以专题的形式进行组织，包括建筑研究概论、设计研究、科学研究、历史理论研究4个模块。学生通过各模块的学习可以较为全面地了解建筑学科内主要的研究领域及相应的思维方式和研究方法。

课程要求
　　将介绍建筑学科的主要研究领域和当代研究前沿，介绍"研究"的特性、思维方式、主要任务、研究的工作架构以及什么是好的研究，帮助学生建立对"研究"的基本认识；介绍文献检索和文献综述的规范和方法；介绍常见的定量研究、定性研究和设计研究的工作方法以及相应的写作规范。

课程内容
1. 综述
2. 文献
3. 科学研究及其方法
4. 科学研究及其写作规范
5. 历史理论研究及其方法
6. 历史理论研究及其写作规范
7. 设计研究及其方法
8. 城市规划理论概述

Teaching objectives
It is a compulsory course to MA. It comprehensively introduces features, ways of thinking and common methods of academic research, and necessary manners of working and writing standard for launching academic research to students. Considering differences of research methods among different fields, teaching and assignment of the course will be organized in the form of special topics, including four parts: introduction to architectural study, design study, scientific study and historical theory study. Through the learning of all parts, students can comprehensively understand main research fields and corresponding ways of thinking and research methods of architecture.

Course requirements
The course introduces main research fields and contemporary research frontiers of architecture, features, ways of thinking, main tasks of "research", working structure of research, and definition of good research to help students form basic understanding of "research". The course also introduces standards and methods of literature retrieval and review, and working methods of common quantitative research, qualitative research and design research, and their corresponding writing standards.

Course contents
1. Review
2. Literature
3. Scientific research and methods
4. Scientific research and writing standards
5. Historical theory study and methods
6. Historical theory study and writing standards
7. Design research and methods
8. Overview of urban planning theory

城市理论课程
URBAN THEORY COURSES

本科一年级
建成环境导论与学科前沿 · 丁沃沃
课程类型：核心
学时/学分：32学时/2学分

Undergraduate Program 1st Year
INTRODUCTION TO ARCHITECTURAL ENVIRONMENT AND FRONTIERS OF DISCIPLINES • DING Wowo
Type: Core
Study Period and Credits: 32 hours / 2 credits

课程介绍
建成环境是人类生产与生活的基本场所，是生存和发展的重要环境，建成环境的优劣关系到每个人的生存状况。建成环境的构建涉及多个学科，与其相关的各类知识是多个学科的共同基础。有史以来，一方面人们依靠技术进步不断从地球获取资源的同时也不断创造出更加高效和舒适的生存空间；另一方面，随着对自然界认知的更新，人们也在不断调整建构建成环境的方法和路径。
因此，从专业的角度了解建成环境的概念、进展、问题和前景对于刚刚踏入学科的初学者来说是后续学习的基础知识。此外，本课程力图承担训练大学生学习方法的任务，以教学过程为载体，引导学生如何借助新的媒体技术获取知识，培养独立思考、思辨的能力，促进学生尽快完成从中学到大学学习方法的转型，为今后的学习打好基础。

课程要求
1. 理解随着社会转型，城市建筑的基本概念在建筑学核心理论中的地位以及认知的视角。
2. 通过理论的研读和案例分析理解建筑形式语言的成因和逻辑，并厘清中、西不同的发展脉络。
3. 通过研究案例的解析理解建筑形式语言的操作并掌握设计研究的方法。

Course descriptions
As an essential place for human production and living and an important environment for their survival and development, built environment matters to everyone's condition. Its construction is a multidisciplinary process, and different varieties of knowledge related are the common foundation of these disciplines. Throughout the history, on the one hand, people rely on technological progress to get resources from the Earth and constantly create a more efficient and comfortable living space; on the other hand, as people update their understanding of nature, they also keep adjusting the methods and paths of building built environment.
Thus, understanding the overview, progress, problems and prospects of built environment from a professional point of view is basic to the continuation for beginners in this course. In addition, this course is designed to train the ways of study in university, taking the Teaching progress as a media, it guides students to acquire knowledge with new media technologies, develops their capacities of independent and critical thinking, and promotes them to switch from a high school learning style to a university learning style as soon as possible, providing a foundation for future study.

Course requirements
1. To understand the position and cognitive perspective of basic concepts of urban architecture in the core theory of architecture with social transformation.
2. To understand the cause and logic of architectural formal languages and different evolution in China and the West through theory study and case analysis.
3. To understand the operation of architectural form languages and master the methods of design research through the analysis of research cases.

本科四年级
城市设计及其理论 · 胡友培
课程类型：选修
学时/学分：36学时/2学分

Undergraduate Program 4th Year
URBAN DESIGN AND THEORY • HU Youpei
Type: Elective Course
Study Period and Credits: 36 hours / 2 credits

课程内容
第一讲：课程概述
第二讲：城市设计技术术语
第三讲：城市设计方法——文本分析：城市设计上位规划；城市设计相关文献；文献分析方法
第四讲：城市设计方法——数据分析：人口数据分析与配置；交通流量数据分析；功能分配数据分析；视线与高度数据分析；城市空间数据模型的建构
第五讲：城市设计方法——城市肌理分类；城市肌理分类概述；肌理形态与建筑容量；肌理形态与开放空间；肌理形态与交通流量；城市绿地指标体系
第六讲：城市设计方法——城市路网组织：城市道路结构与交通结构概述；城市路网与城市功能；城市路网与城市空间；城市路网与市政设施；城市道路断面设计
第七讲：城市设计方法——城市设计表现：城市设计分析图；城市设计概念表达；城市设计成果解析图；城市设计地块深化设计表达；城市设计空间表达
第八讲：城市设计的历史与理论
第九讲：城市路网形态
第十讲：城市空间
第十一讲：城市形态学
第十二讲：城市形态的物理环境
第十三讲：景观都市主义
第十四讲：城市自组织现象及其研究
第十五讲：建筑学图式理论与方法
第十六讲：课程总结

Course contents
Lecture 1: Course overview
Lecture 2: Technical terminology of urban design
Lecture 3: Urban design methods—Text analysis: Urban design upper level planning; relevant literature on urban design; literature analysis methods
Lecture 4: Urban design methods—Data analysis: Population data analysis and configuration; traffic flow data analysis; functional allocation data analysis; analysis of the line of sight and height data; construction of urban spatial data model
Lecture 5: Urban design methods—Urban texture classification: Overview of urban texture classification; texture morphology and building capacity; texture morphology and open space; texture morphology and traffic flow; index system of urban green space
Lecture 6: Urban design methods—Urban road network organization: Overview of urban road structure and traffic structure; urban road network and urban functions; urban road network and urban space; urban road network and municipal facilities; urban road section design
Lecture 7: Urban design methods—Urban design performance: Urban design analysis charts; urban design concept expression; analysis diagram of urban design achievements; deepening the design expression of urban design plots; urban design space expression
Lecture 8: History and theory of urban design
Lecture 9: Urban road network form
Lecture 10: Urban space
Lecture 11: Urban morphology
Lecture 12: The physical environment of urban form
Lecture 13: Landscape urbanism
Lecture 14: Urban self organization phenomenon and its research
Lecture 15: Architectural schema theory and methods
Lecture 16: Course summary

本科四年级
景观规划设计及其理论 · 尹航
课程类型：选修
学时 / 学分：36 学时 / 2 学分

Undergraduate Program 4th Year
LANDSCAPE PALNNING DESIGN AND THEORY • YIN Hang
Type: Elective Course
Study Period and Credits: 36 hours / 2 credits

课程介绍
景观规划设计的对象包括所有的室外环境，景观与建筑的关系往往是紧密而互相影响的，这种关系在城市中表现得尤为明显。景观规划设计及其理论课程希望从景观设计理念、场地设计技术和建筑周边环境塑造等方面开展课程的教学，为建筑学本科生建立更加全面的景观知识体系，并且完善建筑学本科生在建筑场地设计、总平面规划与城市设计等方面的设计能力。

本课程主要从三个方面展开。一是理念与历史；以历史的视角介绍景观学科的发展过程，让学生对景观学科有一个宏观的了解，初步理解景观设计理念的发展。二是场地与文脉；通过阐述景观规划设计与周边自然环境、地理位置、历史文脉和方案可持续性的关系，建立场地与文脉的设计思维。三是景观与建筑；通过设计方法授课、先例分析作业等方式让学生增强建筑的环境意识，了解建筑的场地设计的影响因素、一般步骤与设计方法，并通过与"建筑设计六"和"建筑设计七"的设计任务书相配合的同步课程设计训练来加强学生景观规划设计的能力。

Course descriptions
The objects of landscape planning design include all outdoor environment; the relationship between landscapes and buildings is often close and interactive, which is especially obvious in a city. This course expects to carry out teaching from perspective of landscape design concept, site design technology, building's peripheral environment creation, etc., to establish a more comprehensive landscape knowledge system for the undergraduate students of architecture, and perfect their design ability in building site design, plan planning and urban design and so on.
This course includes three aspects. First, concepts and history: Introduce the development process of landscape discipline from a historical perspective, so that students can have a macro understanding of landscape discipline and preliminarily understand the development of landscape design concepts. The second is the site and context: The design thinking of the site and context is established by explaining the relationship between the landscape planning and design and the surrounding natural environment, geographical locations, historical contexts and program sustainability. Third, landscape and architecture: Through design method teaching, precedent analysis, etc. students can enhance their environmental awareness of architecture, understand the influencing factors, general steps and design methods of building site design, and strengthen their ability of landscape planning and design through synchronous course design training in conjunction with the design tasks of Architectural Design VI and Architectural Design VII.

研究生一年级
城市形态与设计方法论 · 丁沃沃
课程类型：必修
学时 / 学分：36 学时 / 2 学分

Graduate Program 1st Year
URBAN MORPHOLOGY AND DESIGN METHOLOGY • DING Wowo
Type: Required Course
Study Period and Credits: 36 hours / 2 credits

课程介绍
建筑学核心理论包括建筑学的认识论和设计方法论两大部分。建筑设计方法论主要探讨设计的认知规律、形式的逻辑、形式语言类型，以及人的行为、环境特征和建筑材料等客观规律对形式语言的选择和形式逻辑的构成策略。为此，设立了以提升建筑设计方法为目的的关于设计方法论的理论课程，作为建筑设计及其理论硕士学位的核心课程。

课程要求
1. 理解随着社会转型，城市建筑的基本概念在建筑学核心理论中的地位以及认知的视角。
2. 通过理论的研读和案例分析理解建筑形式语言的成因和逻辑，并厘清中、西不同的发展脉络。
3. 通过研究案例的解析理解建筑形式语言的操作并掌握设计研究的方法。

课程内容
第一讲：序言
第二讲：西方建筑学的基础
第三讲：中国——建筑的意义
第四讲：背景与文献研讨
第五讲：历史观与现代性
第六讲：现代城市形态演变与解析
第七讲：现代城市的"乌托邦"
第八讲：现代建筑的意义
第九讲：建筑形式的反思与探索
第十讲：建筑的量产与城市问题
第十一讲："乌托邦"的实践与反思
第十二讲：都市实践探索的理论价值
第十三讲：城市形态的研究
第十四讲：城市空间形态研究的方法
第十五讲：回归理性——建筑学方法论的新进展
第十六讲：建筑学与设计研究的意义
第十七讲：结语与研讨（一）
第十八讲：结语与研讨（二）

Course descriptions
Core theory of architecture includes epistemology and design methodology of architecture. Architectural design methodology mainly discusses cognitive laws of design, logic of forms and types of formal language, and the choice of formal language from objective laws such as human behaviors, environmental features and building materials, and composition strategies of formal logic. Thus, the theory course about design methodology to promote architectural design methods is established as the core course for masters of architectural design and theory.

Course requirements
1. To understand the status and cognitive perspective of basic concept of urban buildings in the core theory of architecture with the social transformation.
2. To understand the reason and logic of architectural formal language and different development processes in China and the West through theory study and case analysis.
3. To understand the operation of architectural formal language and grasp methods of design and study by analyzing cases.

Course contents
Lecture 1: Introduction
Lecture 2: Foundation of western architecture
Lecture 3: China—Meaning of architecture
Lecture 4: Background and literature discussion
Lecture 5: Historicism and modernity
Lecture 6: Analysis and morphological evolution of modern cities
Lecture 7: "Utopia" of modern cities
Lecture 8: Meaning of modern architecture
Lecture 9: Reflection and exploration of architectural forms
Lecture 10: Mass production of buildings and urban problems
Lecture 11: Practice and reflection of "Utopia"
Lecture 12: Theoretical value of exploration on urban practice
Lecture 13: Study on urban morphology
Lecture 14: Methods of urban spatial morphology study
Lecture 15: Return to rationality—New developments of methodology on architecture
Lecture 16: Meaning of architecture and design study
Lecture 17: Conclusion and discussion (1)
Lecture 18: Conclusion and discussion (2)

研究生一年级
当代景观都市实践 · 华晓宁
课程类型：选修
学时/学分：18学时 / 1学分

Graduate Program 1st Year
CONTEMPORARY LANDSCAPE URBANISM PRACTICE • HUA Xiaoning
Type: Elective Course
Study Period and Credits: 18 hours / 1 credit

课程介绍
本课程作为国内首次以景观都市主义相关理论与策略为教学内容的尝试，介绍了景观都市主义思想产生的背景、缘起及其主要理论观点，并结合实例，重点分析了其在不同的场址和任务导向下发展起来的多样化的实践策略和操作性工具。

课程要求
1. 要求学生了解景观都市主义思想产生的背景、缘起和主要理念。
2. 要求学生能够初步运用景观都市主义的理念和方法分析和解决城市设计问题，从而在未来的城市设计实践中强化景观整合意识。

课程内容
第一讲：从图像到效能——景观都市实践的历史演进与当代视野
第二讲：生态效能导向的景观都市实践（一）
第三讲：生态效能导向的景观都市实践（二）
第四讲：社会效能导向的景观都市实践
第五讲：基础设施景观都市实践
第六讲：当代高密度城市中的地形学
第七讲：城市图绘与图解
第八讲：从原型到系统——AA景观都市主义

Course descriptions
Combining relevant theories and strategies of landscape urbanism firstly in China, the course introduces the background, origin and main theoretical viewpoint of landscape urbanism, and focuses on diversified practical strategies and operational tools developed under different orientations of sites and tasks with examples.

Course requirements
1. Students are required to understand the background, origin and main concept of landscape urbanism.
2. Students are required to preliminarily utilize the concept and method of landscape urbanism to analyze and solve problems of urban design, so as to strengthen landscape integration consciousness in the future.

Course contents
Lecture 1: From pattern to performance—Historical revolution and contemporary view of practice of landscape urbanism
Lecture 2: Eco-efficiency-oriented practice of landscape urbanism (1)
Lecture 3: Eco-efficiency-oriented practice of landscape urbanism (2)
Lecture 4: Social efficiency-oriented practice of landscape urbanism
Lecture 5: Infrastructure practice in landscape urbanism
Lecture 6: Geomorphology in contemporary high-density cities
Lecture 7: Urban painting and diagrammatizing
Lecture 8: From the prototype to the system—AA landscape urbanism

---

研究生一年级
城市形态学 · 谷凯
课程类型：选修
学时/学分：36学时 / 2学分

Graduate Program 1st Year
URBAN MORPHOLOGY • GU Kai
Type: Elective Course
Study Period and Credits: 36 hours / 2 credits

课程介绍
本课程介绍城市形态学是一个研究领域以及规划和城市设计的基础。在回顾城市形态学的起源和发展的基础上，本课程鼓励学生探索在城市形态学方面的研究兴趣，并与其他学生分享他们的兴趣、知识和经验。本课程重点关注城市形态理论、规划和城市设计实践以及中国城市发展问题之间现有与潜在的联系，希望学生能够更广泛地了解城市形态领域及其在研究和实践中的应用。

教学模式
课程包括两个主要部分：介绍城市形态研究与实践的讲座和学生之间进行学术交流的研讨会。在确定与其研究兴趣相关的理论和分析工具的基础上，学生将探索城市形态学在中国背景下的应用。为了使课堂成为一个有效的共同学习场所，所有参与者都必须充分参与（而不仅仅是旁观者），并愿意以建设性的方式提出不同意见，这样才能更好地进行讨论，或使重要的观点和差异得以显现。

Course descriptions
This course introduces urban morphology as a field of research and as a basis for planning and urban design. Based on a review of the origin and development of urban morphology, this course encourages students to explore their own research interests in urban morphology and to share their interest, knowledge and experience with other students. Focusing on the existing and potential linkages between urban morphological theories, planning and urban design practice and urban development issues in China, students are expected to gain a wider understanding of the field of urban morphology and its use in research and practice.

Learning modes
The course has two major components: lectures introducing urban morphological research and practice, seminars providing scholarly interchange among students. Based on the identification of theories and analytical tools related to their research interests, students will explore the application of urban morphology in the context of China. To make the classroom an effective place for learning together, it is important for all participants to be fully involved (not simply spectators) and to be willing to give opinions constructively where this will lead to a better discussion or enable important ideas and differences in perspective to emerge.

历史理论课程
HISTORY THEORY COURSES

本科一年级
建筑通史 · 王骏阳
课程类型：平台（工科试验班）
学时 / 学分：36 学时 / 2 学分

Undergraduate Program 1st Year
GENERAL HISTORY OF ARCHITECTURE ·
WANG Junyang
Type: Platform (Engineering Experimental Class)
Study Period and Credits: 36 hours / 2 credits

教学目标
1 打破数十年以来建筑史教学中中外之间的学科壁垒，建立人类建筑文化之间的相互联系和全球视野中的建筑史认识。
2. 着重古今中外的融会贯通，弥补网络时代知识碎片化的不足。
3. 贯通建筑史教学与建筑学概论教学，通过历史的视角建立健全学科基础认知。
4. 将建筑史专业知识与人类文明普遍认知相结合，使专业基础认知课具有更为广泛的通识选修课的可能和吸引力。

课程内容
第一讲：走出建筑与建筑物的悖论
第二讲：居住的起源与发展（一）
第三讲：居住的起源与发展（二）
第四讲：神秘的人类早期文明
第五讲：融合与冲突中的西亚与伊斯兰建筑
第六讲：古希腊和古罗马建筑
第七讲：欧洲中世纪建筑
第八讲：意大利文艺复兴建筑及后续
第九讲：现代建筑
第十讲：后现代建筑
第十一讲：装饰的问题
第十二讲：空间、结构、形式
第十三讲：日本建筑
第十四讲：复习题讲解
第十五讲：复习题讲解
第十六讲：世界建筑史中的中国建筑

Teaching objectives
1. Break the disciplinary barrier that has existed in the teaching of architectural history for decades between China and foreign countries, and establishing the interconnection among human architectural cultures and a global understanding of the architectural history.
2. Emphasize the integration between ancient and modern, Chinese and foreign cultures to make up for the shortcomings of knowledge fragmentation in the Internet era.
3. Integrate the teaching of Architectural History and the teaching of Introduction to Architecture, establish and improve a sound disciplinary foundation through a historical perspective.
4. Combine the professional knowledge of architectural history with the general understanding of human civilization, making the professional basic cognitive courses more likely become general elective courses.

Course contents
Lecture 1: Breaking Out of the Paradox between Architecture and Buildings
Lecture 2: The Origin and Development of Residence (1)
Lecture 3: The Origin and Development of Residence (2)
Lecture 4: Mysterious Early Human Civilization
Lecture 5: Western Asia and Islamic architecture in Integration and Conflict
Lecture 6: Ancient Greek and Ancient Roman architecture
Lecture 7: Medieval European Architecture
Lecture 8: Italian Renaissance architecture and Follow up
Lecture 9: Modern Architecture
Lecture 10: Postmodern Architecture
Lecture 11: Decoration Issues
Lecture 12: Space, Structure, and Form
Lecture 13: Japanese Architecture
Lecture 14: Explanation of Exercises
Lecture 15: Explanation of Exercises
Lecture 16: Chinese Architecture in World Architecture History

---

本科二年级
中国建筑史（古代） · 赵辰 史文娟 赵潇欣
课程类型：必修
学时 / 学分：32 学时 / 2 学分

Undergraduate Program 2nd Year
HISTORY OF CHINESE ARCHITECTURE (ANCIENT) · ZHAO Chen, SHI Wenjuan, ZHAO Xiaoxin
Type: Required Course
Study Period and Credits: 32 hours / 2 credits

课程内容
课程以系列讲座的形式展开三大专题，精心设计了三大环环相扣作业。
1. "建构文化"专题，论述以木构框架与土构围合为基本特征单体建筑营造，配合"中国房子"的作业——提供四个典型地域的基本建造体系的信息，及建筑基本单元平立剖图纸。通过数字三维模型，模拟营造系统的全过程，了解传统建造材料与工序，并通过不同地域迥异的建造方式与最终建筑形式的差异，引导学生切实理解讲座内容所阐释的建筑形式与建造体系之间的必然关系。
2. "人居文化"专题，讲授以院落为其核心建筑群体空间组织，配合"中国院子"的作业——提供江南某历史县城一街廓内的基地及相关建筑要素模型。要求学生首先策划日常生活起居过程，进行传统生活起居行为模式的"脚本"写作；随后在基地上，依据地界、以给定的建筑要素，遵照前期自行设定的"脚本"，组合构建院落的三维模型体。
3. "城镇文化"专题，讲述聚落空间与城市形态，配合"中国园子"的作业——延续前期院子的脚本，进行传统园居行为模式的"脚本"写作，随后在宅旁基地上，依据地界，结合基地周遭环境及已有的"院子"布局，遵照前期自行设计以给定的园林要素，完成园林的布局并表现。
4. 深层次理解中国传统的人居空间——建构文化的作业设置，令同学们在模拟营造的过程中，深刻认识到传统的"土木 / 营造"体系所提供的物质性空间，同与之相适应的人居行为为习俗，共同形塑了传统人居形态。

Course contents
The course presents three major topics in the form of a series of lectures, and meticulously designs three interconnected assignments.
1. "Constructing Culture" topic discusses the construction of individual buildings with the basic characteristics of wooden frames and soil structures, in conjunction with the operation of "Chinese houses"—Providing information on the basic construction systems of four typical regions, as well as basic single element plan and elevation drawings of buildings. Through digital 3D models, simulate the entire process of the construction a system, understand traditional construction materials and processes, and guide students to truly understand the inevitable relationship between building forms and construction systems explained in the lecture content through the differences in construction methods and final building forms in different regions.
2. "Human Settlement Culture" topic teaches the spatial organization which takes the courtyard as its core architectural group, in conjunction with the operation of "Chinese Courtyards"—Providing a base and related architectural element models within a street boundary of a historical county in Jiangnan. Students are required to first plan the daily living process and write "scripts" of traditional living behavior patterns; Subsequently, on the house base, based on the boundaries and given architectural elements, a three-dimensional model of the courtyard is constructed by following the "script" set in the previous stage.
3. "Urban Culture" topic discusses the relationship between settlement space and urban forms, in conjunction with the assignment of "Chinese Gardens"—Continuing the script of the previous courtyard, writing the "script" of traditional garden living behavior patterns. Then, on the homestead site, based on the boundaries, combined with the surrounding environment of the site and the existing "courtyard" layout, the layout and performance of the garden should be completed according to the given garden elements designed in the early stage.
4. Deeply understanding the traditional Chinese living space—The homework setting of constructing culture enables students to deeply understand the physical space provided by the traditional "civil engineering/construction" system during the simulation process, and the corresponding living behaviors and customs, which together shape the traditional living form.

本科二年级
外国建筑史（古代）・王骏阳
课程类型：必修
学时 / 学分：36 学时 /2 学分

Undergraduate Program 2nd Year
HISTORY OF WESTERN ARCHITECTURE (ANCIENT) • WANG Junyang
Type: Required Course
Study Period and Credits: 36 hours / 2 credits

教学目标
　　本课程力图对西方建筑史的脉络做一个整体勾勒，使学生在掌握重要的建筑史知识点的同时，对西方建筑史在 2 000 多年里的变迁的结构转折（不同风格的演变）有深入的理解。本课程希望学生对建筑史的发展与人类文明发展之间的密切关联有所认识。

课程内容
1. 概论　2. 希腊建筑　3. 罗马建筑　4. 中世纪建筑
5. 意大利的中世纪建筑　6. 文艺复兴　7. 巴洛克
8. 美国城市　9. 北欧浪漫主义
10. 加泰罗尼亚建筑　11. 先锋派
12. 德意志制造联盟与包豪斯　13. 苏维埃的建筑与城市
14. 1960 年代的建筑　15. 1970 年代的建筑
16. 答疑

Teaching objectives
This course seeks to give an overall outline of Western architectural history, enables students to master important knowledge points of architectural history, and may have an in-depth understanding of the structural transition (evolution of different styles) of Western architectural history in over 2 000 years. This course hopes that students can understand the close association between the development of architectural history and the development of human civilization.

Course contents
1. Generality　2. Greek architecture　3. Roman architecture
4. The Middle Ages architecture 5. The Middle Ages architecture in Italy　6. Renaissance 7. Baroque　8. American cities
9. Nordic romanticism 10. Catalonian architecture　11. Avant-garde
12. German manufacturing alliance and Bauhaus
13. Soviet architecture and cities　14. 1960's architecture
15. 1970's architecture　16. Answer questions

本科三年级
外国建筑史（当代）・胡恒
课程类型：必修
学时 / 学分：36 学时 /2 学分

Undergraduate Program 3rd Year
HISTORY OF WESTERN ARCHITECTURE (MODERN) • HU Heng
Type: Required Course
Study Period and Credits: 36 hours / 2 credits

教学目标
　　本课程力图用专题的方式对文艺复兴时期的 7 位代表性的建筑师与 5 位现当代的重要建筑师作品做一细致的讲解。本课程将重要建筑师的全部作品尽可能在课程中梳理一遍，使学生能够全面掌握重要建筑师的设计思想、理论主旨、与时代的特殊关联、在建筑史中的意义。

课程内容
1. 伯鲁乃列斯基　2. 阿尔伯蒂 3. 伯拉孟特
4. 米开朗琪罗（1）5. 米开朗琪罗（2）
6. 罗马诺　7. 桑索维诺
8. 帕拉蒂奥（1）　9. 帕拉蒂奥（2）
10. 赖特　11. 密斯
12. 勒·柯布西耶（1）13. 勒·柯布西耶（2）
14. 海杜克　15. 妹岛和世
16. 答疑

Teaching objectives
This course seeks to make a detailed explanation to the works of 7 representative architects in the Renaissance period and 5 important modern and contemporary architects in a special topic way. This course will try to introduce all works of these important architects, so that the students can fully grasp their design ideas, theoretical subjects, their particular relevance with the era and the significance in the architectural history.

Course contents
1. Brunelleschi　2. Alberti　3. Bramante
4. Michelangelo(1)　5. Michelangelo(2)
6. Romano　7. Sansovino　8. Palladio(1)　9. Palladio(2)
10. Wright　11. Mies 12. Le Corbusier(1)　13. Le Corbusier(2)
14. Hejduk　15. Kazuyo Sejima
16. Answer questions

本科三年级
中国建筑史（近现代）・赵辰　冷天
课程类型：必修
学时 / 学分：36 学时 /2 学分

Undergraduate Program 3rd Year
HISTORY OF CHINESE ARCHITECTURE (MODERN) • ZHAO Chen, LENG Tian
Type: Required Course
Study Period and Credits: 36 hours / 2 credits

课程介绍
　　本课程作为本科建筑学专业的历史与理论课程，是中国建筑史教学中的一部分。在中国与西方的古代建筑历史课程的基础上，了解中国社会进入近代，以至于现当代的发展进程。
　　在对比中西方建筑文化的基础之上，建立对中国近现代建筑的整体认识。深刻理解中国传统建筑文化在近代以来与西方建筑文化的冲突与相融之下，逐步演变发展至今天成为世界建筑文化的一部分之意义。

Course descriptions
As the history and theory course for undergraduate students of architecture, this course is part of the teaching of history of Chinese architecture. Based on the earlier studying of Chinese and western history of ancient architecture, students can understand the development process as Chinese society steps into modern times and even the contemporary age.
Based on the comparison between Chinese and western building culture, establish the overall understanding of China's modern and contemporary buildings. Have further understanding of the significance of China's traditional building culture's gradual evolution into one part of today's world building culture under the conflict and blending with Western building culture in modern times.

研究生一年级
建筑理论研究・赵辰
课程类型：必修
学时 / 学分：18 学时 /1 学分

Graduate Program 1st Year
STUDIES OF ARCHITECTURAL THEORY • ZHAO Chen
Type: Required Course
Study Period and Credits: 18 hours / 1 credit

课程介绍
　　了解中、西方学者对中国建筑文化诠释的发展过程，理解新的建筑理论体系中对中国建筑文化重新诠释的必要性，学习重新诠释中国建筑文化的建筑观念与方法。

课程内容
1. 本课的总览和基础
2. 中国建筑：西方人的诠释与西方建筑观念的改变
3. 中国建筑：中国人的诠释以及中国建筑学术体系的建立
4. 木结构体系：中国建构文化的意义
5. 住宅与园林：中国人居文化意义
6. 宇宙观的和谐：中国城市文化的意义
7. 讨论

Course descriptions
Understand the development process of Chinese and western scholars' interpretation of Chinese architectural culture, understand the necessity of reinterpretation of Chinese architectural culture in the new architectural theory system, and learn the architectural concepts and methods of reinterpretation of Chinese architectural culture.

Course contents
1. Overview and foundation of this course
2. Chinese architecture: Western interpretation and the change of western architectural concepts
3. Chinese architecture: Chinese interpretation and the establishment of Chinese architecture academic system
4. Wood structure system: The significance of Chinese construction culture
5. Residences and gardens: The cultural significance of human settlement in China
6. Harmony of cosmology: The significance of Chinese urban culture
7. Discussion

研究生一年级
建筑理论研究 · 王骏阳
课程类型：必修
学时/学分：18学时/1学分

Graduate Program 1st Year
STUDIES OF ARCHITECTURAL THEORY ·
WANG Junyang
Type: Required Course
Study Period and Credits: 18 hours / 1 credit

课程介绍
本课程是针对研究生的西方建筑史教学的一部分。主要涉及当代西方建筑界具有代表性的思想和理论，其主题包括历史主义、先锋建筑、批判理论、建构文化以及对当代城市的解读等。本课程大量运用图片资料，广泛涉及哲学、历史、艺术等领域，力求在西方文化发展的背景中呈现建筑思想和理论的相对独立性及关联性，理解建筑作为一种人类活动所具有的社会和文化意义，启发学生的理论思维和批判精神。

课程内容
第一讲：建筑理论概论
第二讲：数字化建筑与传统建筑学的分离与融合
第三讲：语言、图解、空间内容
第四讲："拼贴城市"与城市的观念
第五讲：建构与营造
第六讲：手法主义与当代建筑
第七讲：从主线历史走向多元历史之后的思考
第八讲：讨论

Course descriptions
This course is a part of western architectural history teaching for graduate students. It mainly deals with the representative thoughts and theories in western architectural circles, including historicism, vanguard architecture, critical theory, tectonic culture and interpretation of contemporary cities etc.. Using a lot of pictures which involve extensive fields including philosophy, history, art, etc., this course attempts to show the relative independence and relevance of architectural thoughts and theories under the development background of western culture, understand the social and cultural significance owned by architecture as human activities, and inspire students' theoretical thinking and critical spirit.

Course contents
Lecture 1: Overview of architectural theories
Lecture 2: Separation and integration between digital architecture and traditional architecture
Lecture 3: Language, diagram and spatial content
Lecture 4: "Collage city" and concept of cities
Lecture 5: Tectonics and Yingzao (Ying-Tsao)
Lecture 6: Mannerism and modern architecture
Lecture 7: Thinking after main-line history to diverse history
Lecture 8: Discussion

研究生一年级
建筑史方法 · 胡恒
课程类型：选修
学时/学分：18学时/1学分

Graduate Program 1st Year
METHOD OF ARCHITECTURAL HISTORY
· HU Heng
Type: Elective Course
Study Period and Credits: 18 hours / 1 credit

教学目标
促进学生对历史研究的主题、方法、路径有初步的认识，通过具体的案例讲解使学生能够理解当代中国建筑史研究的诸多可能性。

课程内容
1. 图像与建筑史研究（1—文学、装置、设计）
2. 图像与建筑史研究（2—文学、装置、设计）
3. 图像与建筑史研究（3—绘画与园林）
4. 图像与建筑史研究（4—绘画、建筑、历史）
5. 图像与建筑史研究（5—文学与空间转译）
6. 方法讨论1
7. 方法讨论2

Teaching objectives
To promote students' preliminary understanding of the topic, method and approach of historical research. To make students understand the possibilities of contemporary study on history of Chinese architecture through explanation for specific cases.

Course contents
1. Images and architectural history study (1–literature, devices and design)
2. Images and architectural history study (2–literature, devices and design)
3. Images and architectural history study (3– painting and gardens)
4. Images and architectural history study (4–painting, architecture and history)
5. Images and architectural history study (5–literature and spatial transform)
6. Method discussion 1
7. Method discussion 2

研究生一年级
中国建构文化（木构）研究 · 赵辰
课程类型：选修
学时/学分：18学时/1学分

Graduate Program 1st Year
STUDIES IN CHINESE WOODEN TECTONIC CULTURE · ZHAO Chen
Type: Elective Course
Study Period and Credits: 18 hours / 1 credit

教学目标
以木为材料的建构文化是世界各文明中的基本成分，中国的木建构文化更是深厚而丰富。在全球可持续发展要求之下，木建构文化必须得到重新的认识和评价。对于中国建筑文化来说，更具有再认识和再发展文化传统的意义。

课程内容
阶段一：理论基础——对全球木建构文化的重新认识
阶段二：中国木建构文化的原则和方法（讲座与工作室）
阶段三：中国木建构的基本形——从家具到建筑（讲座与工作室）
阶段四：结构造型的发展和木建构的现代化（讲座）
阶段五：建造实验的鼓动（讲座与工作室）

Teaching objectives
The wood – based construction culture is the basic component of all civilizations in the world, and Chinese wood construction culture is profound and abundant. Under the requirement of global sustainable development, wood construction culture must be re-recognized and re-evaluated. For Chinese architectural culture, it is of great significance to re-recognize and re-develop the cultural tradition.

Course contents
Stage 1: Theoretical basis—Re-understanding of global wood construction culture
Stage 2: Principles and methods of Chinese wood construction culture (lectures and studios)
Stage 3: The basic shape of Chinese wood construction—From furniture to architecture (lectures and studios)
Stage 4: Development of structural modeling and modernization of wood construction (lectures)
Stage 5: Agitation of construction experiment (lectures and studios)

本科二年级
CAAD 理论与实践（一）· 吉国华 傅筱 万军杰
课程类型：选修
学时 / 学分：36 学时 / 2 学分

Undergraduate Program 2nd Year
THEORY AND PRACTICE OF CAAD 1 • JI Guohua, FU Xiao, WAN Junjie
Type: Elective Course
Study Period and Credits: 36 hours / 2 credits

课程介绍
在现阶段的 CAD 教学中，强调了建筑设计在建筑学教学中的主干地位，将计算机技术定位为绘图工具，本课程就是帮助学生可以尽快并且熟练地掌握如何利用计算机工具进行建筑设计的表达。课程中整合了 CAD 知识、建筑制图知识以及建筑表现知识，从传统 CAD 教学中教会学生用计算机绘图的模式向教会学生用计算机绘制有形式感的建筑图的模式转变，强调准确性和表现力作为评价 CAD 学习的两个最重要指标。
本课程的具体学习内容包括：
1. 初步掌握 AutoCAD 软件和 SketchUP 软件的使用，能够熟练完成二维制图和三维建模的操作；2. 掌握建筑制图的相关知识，包括建筑投影的基本概念，平面图、立面图、剖面图、轴测图、透视图和阴影图的制图方法和技巧；3. 掌握图面效果表达的技巧，包括黑白线条图和彩色图纸的表达方法和排版方法。

Course descriptions
The core position of architectural design is emphasized in the CAD course. The computer technology is defined as a drawing instrument. The course helps students learn how to make architectural presentation using computers fast and expertly. The knowledge of CAD, architectural drawing and architectural presentation are integrated into the course. The traditional mode of teaching students to draw in CAD course will be transformed into teaching students to draw by computers architectural drawing with sense of forms. The precision and expression will be emphasized as two most important factors to estimate the teaching effect of CAD course.
Contents of the course include:
1. Use AutoCAD and SketchUP to achieve the 2D drawing and 3D modeling expertly; 2. Learn relative knowledge of architectural drawing, including basic concepts of architectural projection, drawing methods and skills of plan, elevation, section, axonometry, perspective and shadow; 3. Master skills of presentation, including the methods of expression and typesetting using mono and colorful drawings.

本科三年级
建筑技术（一）· 傅筱 李清朋
课程类型：必修
学时 / 学分：36 学时 / 2 学分

Undergraduate Program 3rd Year
ARCHITECTURAL TECHNOLOGY 1 • FU Xiao, LI Qingpeng
Type: Required Course
Study Period and Credits: 36 hours / 2 credits

课程介绍
本课程是建筑学专业本科生的专业主干课程。本课程的任务主要是以建筑师的工作性质为基础，讨论一个建筑生成过程中最基本的三大技术支撑（结构、构造、施工）的原理性知识要点，以及它们在建筑实践中的相互关系。

Course descriptions
The course is a major course for the undergraduate students of architecture. The main purpose of this course is based on the nature of the architects' work, to discuss the principle knowledge points of the basic three kinds of technical support in the process of generating construction (structure, construction, execution), and their mutual relations in the architectural practice.

本科三年级
建筑技术（二）声光热 · 吴蔚
课程类型：必修
学时 / 学分：36 学时 / 2 学分

Undergraduate Program 3rd Year
ARCHITECTURAL TECHNOLOGY 2 SOUND, LIGHT AND HEAT • WU Wei
Type: Required Course
Study Period and Credits: 36 hours / 2 credits

课程介绍
本课程是针对三年级学生所设计，课程介绍了建筑热工学、建筑光学、建筑声学中的基本概念和基本原理，使学生能掌握建筑的热环境、声环境、光环境的基本评估方法，以及相关的国家标准。学生完成学业后在此方向上能阅读相关书籍，具备在数字技术方法等相关资料的帮助下，完成一定的建筑节能设计的能力。

Course descriptions
Designed for the Grade 3rd students, this course introduces the basic concepts and principles in architectural thermal engineering, architectural optics and architectural acoustics, so that the students can master the basic methods for the assessment of buildings' thermal environment, sound environment and light environment as well as the related national standards. After graduation, students will be able to read the related books regarding these aspects, and have the ability to complete certain building energy efficiency design with the help of the related digital techniques and methods.

本科三年级
建筑技术（三）水电暖 · 吴蔚
课程类型：必修
学时 / 学分：36 学时 / 2 学分

Undergraduate Program 3rd Year
ARCHITECTURAL TECHNOLOGY 3 WATER, ELECTRICITY AND HEATING • WU Wei
Type: Required Course
Study Period and Credits: 36 hours / 2 credits

课程介绍
本课程是针对南京大学建筑与城市规划学院本科三年级学生所设计。课程介绍了建筑给水排水系统、采暖通风与空气调节系统、电气工程的基本理论、基本知识和基本技能，使学生能熟练地阅读水电、暖通工程图，熟悉水电及消防的设计、施工规范，了解燃气供应、安全用电及建筑防火、防雷的初步知识。

Course descriptions
This course is designed for third year undergraduate students in the School of Architecture and Urban Planning, Nanjing University. The course introduces the basic theories, knowledge, and skills of building water supply and drainage systems, heating, ventilation, and air conditioning systems, and electrical engineering, enabling students to proficiently read water, electricity, and HVAC engineering drawings, familiarize themselves with the design and construction specifications of water, electricity, and fire protection, and gain a preliminary understanding of gas supply, safe electricity use, building fire prevention, and lightning protection.

本科三年级
可持续设计与技术 · 梁卫辉
课程类型：选修
学时 / 学分：36 学时 / 2 学分

Undergraduate Program 3rd Year
SUSTAINABLE DESIGN AND TECHNOLOGY ·
Liang Weihui
Type: Elective Course
Study Period and Credits: 36 hours / 2 credits

教学目标
　　本课程的主要任务是使学生建立可持续建筑设计的观念，加深理解建筑节能对实现国家"双碳"目标的重要意义。通过课程教学，使学生了解国内外建筑节能设计和可持续建筑技术方面的发展状况及发展趋势，明确可持续建筑设计的基本原理，掌握节能建筑构造做法以及整体建筑物理环境系统模拟的基本方法，为学生拓宽知识面打下基础。

课程内容
第一章：可持续建筑发展背景
第二章：建筑热环境与建筑节能技术
第三章：建筑能耗模拟
第四章：建筑风环境理论与营造技术
第五章：可再生能源利用与技术
第六章：水资源与材料资源利用技术
第七章：可持续建筑案例

Teaching objectives
The main task of this course is to help students establish a concept of sustainable architectural design and deepen their understanding of the importance of building energy efficiency in achieving the national "dual carbon" goals. Through course teaching, students can understand the development status and trends of energy-saving design and sustainable building technology both domestically and internationally, clarify the basic principles of sustainable building design, master the construction methods of energy-saving buildings and the basic methods of simulating the overall physical environment system of buildings, and lay a foundation to broaden their knowledge.

Course contents
Chapter 1: Background of sustainable building development
Chapter 2: Building thermal environment and building energy conservation technology
Chapter 3: Simulation of building energy consumption
Chapter 4: Architectural wind environment theory and construction technology
Chapter 5: Renewable energy utilization and technology
Chapter 6: Water resources and material resources utilization technology
Chapter 7: Sustainable building cases

本科四年级
建设工程项目管理 · 谢明瑞
课程类型：选修
学时 / 学分：36 学时 / 2 学分

Undergraduate Program 4th Year
MANAGEMENT OF CONSTRUCTION PROJECT · XIE Mingrui
Type: Elective Course
Study Period and Credits: 36 hours / 2 credits

课程介绍
　　帮助学生系统掌握建设工程项目管理的基本概念、理论体系和管理方法，了解建筑规划设计在建设工程项目中的地位、特点和重要性。
　　延展建筑学专业学生基本知识结构层面，拓展学生的发展方向。

Course descriptions
To help students systematically master the basic concept, theoretical system and management method of construction project management, understand the position, characteristics and importance of architectural planning design in the construction project.
To extend the basic knowledge structure level of students majoring in architecture, extend the development direction of students.

本科四年级
建筑学中的技术人文主义 · 窦平平
课程类型：选修
学时 / 学分：36 学时 / 2 学分

Undergraduate Program 4th Year
TECHNOLOGY OF HUMANISM IN ARCHITECTURE · DOU Pingping
Type: Elective Course
Study Period and Credits: 36 hours / 2 credits

课程介绍
　　课程详尽阐释了为满足建筑的多方需求而投入的技术探索和人文关怀。课程包括四大版块，共 16 个主题讲座，以案例精读的形式引介相关建筑师和学者的作品和理论。希望培养学生对建筑学中的技术议题具有批判性和人文主义的深入理解。

Course descriptions
This course elaborates the technological endeavors and humanistic concern in fulfilling the multiaspect architectural demands. It takes shape in a series of sixteen theme lectures, grouped in four sections, introducing prominent architects and scholars through richly illustrated case studies and interpretations. It aims to nurture the students with a critical and humanistic understanding of the role of technology playing in the discipline of architecture.

本科四年级
CAAD 理论与实践（二） · 吉国华
课程类型：选修
学时 / 学分：18 学时 / 1 学分

Undergraduate Program 4th Year
THEORY AND PRACTICE OF CAAD · JI Guohua
Type: Elective Course
Study Period and Credits: 18 hours / 1 credit

教学目标
　　随着计算机辅助建筑设计技术的快速发展，当前数字技术在建筑设计中的角色逐渐从辅助绘图转向了真正的辅助设计，并引发了设计的革命和建筑的形式创新。本课程讲授 Grasshopper 参数化编程建模方法以及相关的几何知识，让学生在掌握参数化编程建模技术的同时，增强以理性的过程思维方式分析和解决设计问题的能力，为数字建筑设计和数字建造打下必要的基础。

大纲内容
　　课程讲授基于 Rhinoceros 的算法编程平台 Grasshopper 的参数化编程建模方法，包括各类运算器的功能与使用、图形的生成与分析、数据的结构与组织、各类建模的思路与方法，以及相应的数学与计算机编程知识。

Teaching objectives
With the rapid development of computer-aided architectural design technology, the role of digital technology in architectural design has gradually shifted from auxiliary drawing to truly auxiliary design, and has triggered a revolution in design and innovation in architectural forms. This course teaches Grasshopper parametric programming modeling methods and related geometric knowledge, enabling students to master parametric programming modeling techniques while enhancing their ability to analyze and solve design problems in a rational process thinking manner, laying a necessary foundation for digital building design and construction.

Outline contents
The course teaches the parametric modeling methods of Grasshopper, an algorithm programming platform based on Rhinoceros, including the functions and usage of various algorithms, graph generation and analysis, data structure and organization, various modeling ideas and methods, as well as corresponding mathematical and computer programming knowledge.

研究生一年级
传热学与计算流体力学基础·郜志
课程类型：选修
学时/学分：36学时/2学分

Graduate Program 1st Year
FUNDAMENTALS OF HEAT TRANSFER AND COMPUTATIONAL FLUID DYNAMICS • GAO Zhi
Type: Elective Course
Study Period and Credits: 36 hours / 2 credits

课程介绍
本课程的主要任务是使建筑学/建筑技术学专业的学生掌握传热学和计算流体力学的基本概念和基础知识，通过课程教学，使学生熟悉传热学中导热、对流和辐射的经典理论，并了解传热学和计算流体力学的最新研究进展，为建筑能源和环境系统的计算和模拟打下坚实的理论基础。教学中尽量简化传热学和计算流体力学经典课程中冗杂公式的推导过程，而着重于如何解决建筑能源与建筑环境中涉及流体流动和传热的实际应用问题。

Course descriptions
This course introduces students majoring in Building Science and Engineering / Building Technology to the fundamentals of heat transfer and computational fluid dynamics (CFD). Students will study classical theories of conduction, convection and radiation when heat transfers, and learn advanced research developments of heat transfer and CFD. The complex mathematics and physics equations are not emphasized. It is desirable that for real-case scenarios students will have the ability to analyze flow and heat transfer phenomena in building energy and environment systems.

研究生一年级
GIS基础与应用·童滋雨
课程类型：选修
学时/学分：18学时/1学分

Graduate Program 1st Year
CONCEPTS AND APPLICATION OF GIS • TONG Ziyu
Type: Elective Course
Study Period and Credits: 18 hours / 1 credit

课程介绍
本课程的主要目的是让学生理解GIS的相关概念以及GIS对城市研究的意义，并能够利用GIS软件对城市进行分析和研究。

Course descriptions
This course aims to enable students to understand the related concept of GIS and the significance of GIS to urban research, and to be able to use GIS software to carry out urban analysis and research.

研究生一年级
材料与建造·冯金龙
课程类型：选修
学时/学分：18学时/1学分

Graduate Program 1st Year
MATERIALS AND CONSTRUCTION • FENG Jinlong
Type: Elective Course
Study Period and Credits: 18 hours / 1 credit

课程介绍
本课程将介绍现代建筑技术的发展过程，论述现代建筑技术及其美学观念对建筑设计的重要作用；探讨由材料、结构和构造方式所形成的建筑建造的逻辑方式；研究建筑形式产生的物质技术基础，诠释现代建筑的建构理论与研究方法。

Course descriptions
It introduces the development process of modern architecture technology and discusses the important role played by the modern architecture technology and its aesthetic concept in the architectural design. It explores the logical methods of architectural construction formed by material, structure and construction. It studies the material and technical basis for the creation of architectural form, and interprets the construction theory and research method for modern architecture.

研究生一年级
计算机辅助技术·吉国华
课程类型：必修
学时/学分：18学时/1学分

Graduate Program 1st Year
COMPUTER AIDED DESIGN • JI Guohua
Type: Required Course
Study Period and Credits: 18 hours / 1 credit

课程介绍
随着计算机辅助建筑设计技术的快速发展，当前数字技术在建筑设计中的角色逐渐从辅助绘图转向了真正的辅助设计，并引发了设计的革命和建筑的形式创新。本课程讲授Grasshopper参数化编程建模方法以及相关的几何学知识，让学生在掌握参数化编程建模技术的同时，增强以理性的过程思维方式分析和解决设计问题的能力，为数字建筑设计和数字建造打下必要的基础。
基于Rhinoceros的算法编程平台Grasshopper的参数化建模方法，讲授内容包括各类运算器的功能与使用、图形的生成与分析、数据的结构与组织、各类建模的思路与方法，以及相应的数学与计算机编程知识。

Course descriptions
With the rapid development of computer-aided architectural design technology, the role of digital technology in architectural design has gradually shifted from auxiliary drawing to truly auxiliary design, and has triggered a revolution in design and innovation in architectural form. The course introduces methods of Grasshopper parametric programming and modeling and relevant geometric knowledge. The course allows students to master these methods, and enhance their ability to analyze and solve designing problems with rational thinking, building necessary foundation for digital architecture design and digital construction.
In this course, the teacher will teach parametric modeling methods based on Grasshopper, an algorithmic programming platform based on Rhinoceros, including functions and application of all kinds of arithmetic units, pattern formation and analysis, structure and organization of data, various thoughts and methods of modeling, and related knowledge of mathematics and computer programming.

研究生一年级
建筑体系整合 · 吴蔚
课程类型：必修 / 选修
学时 / 学分：18~36 学时 /1~2 学分

Graduate Program 1st Year
BUILDING SYSYTEM INTEGRATION · WU Wei
Type: Required/Elective Course
Study Period and Credits: 18-36 hours / 1-2 credits

课程介绍
本课程是从建筑各个体系整合的角度来解析建筑设计。首先，课程介绍了建筑体系整合的基本概念、原理及其美学观念；然后具体解读以上各个设计元素在整个建筑体系中所扮演的角色及其影响力，了解建筑各个系统之间的互相联系和作用；最后，以全球的环境问题和人类生存与发展为着眼点，引导同学们重新审视和评判我们奉为信条的设计理念和价值系统。本课程着重强调建筑设计需要了解不同学科和领域的知识，熟悉各工种之间的配合和协调。

Course descriptions
A building is an assemblage of materials and components to obtain a shelter from external environment with a certain amount of safety so as to provide a suitable internal environment for physiological and psychological comfort in an economical manner. This course examines the role of building technology in architectural design, shows how environmental concerns have shaped the nature of buildings, and takes a holistic view to understand the integration of different building systems. It employs total building performance which is a systematic approach, to evaluate the performance of various sub-systems and to appraise the degree of integration of the sub-systems.

研究生一年级
算法设计 · 吉国华 童滋雨
课程类型：选修
学时 / 学分：36 学时 / 2 学分

Graduate Program 1st Year
ALGORITHM DESIGN · JI Guohua TONG Ziyu
Type: Elective Course
Study Period and Credits: 36 hours / 2 credits

课程介绍
编程技术是数字建筑的基础，本课程主要讲授 Grasshopper 脚本编程和 Processing 编程，让学生在掌握代码编程基础技术的同时，增强以理性的过程思维方式分析和解决设计问题的能力，逐步掌握数字设计的方法，为数字设计和建造课程打好基础。

Course descriptions
Programming technology is the foundation of digital architecture, this course mainly teaches Grasshopper script programming and Processing programming, so that students can master the basic technology of code programming, at the same time, enhance the their ability to analyze and solve design problems with rational process thinking, gradually master the method of digital design, and lay a good foundation for the course of digital design and construction.

研究生一年级
建设工程项目管理 · 谢明瑞
课程类型：选修
学时 / 学分：36 学时 / 2 学分

Graduate Program 1st Year
MANAGEMENT OF CONSTRUCTION PROJECT · XIE Mingrui
Type: Elective Course
Study Period and Credits: 36 hours / 2 credits

课程介绍
帮助学生系统掌握建设工程项目管理的基本概念、理论体系和管理方法，了解建筑规划设计在建设工程项目中的地位、特点和重要性。
延展建筑学专业学生基本知识结构层面，拓展学生的发展方向。

Course descriptions
To help students systematically master the basic concept, theoretical system and management method of construction project management, understand the position, characteristics and importance of architectural planning design in the construction project.
To extend the basic knowledge structure level of students majoring in architecture, extend the development direction of students.

研究生一年级
建筑环境学与设计 · 尤伟 郜志
课程类型：必修
学时 / 学分：18 学时 / 1 学分

Graduate Program 1st Year
ARCHITECTURAL ENVIRONMENTAL SCIENCE AND DESIGN · YOU Wei, GAO Zhi
Type: Required Course
Study Period and Credits: 18 hours / 1 credit

课程介绍
本课程是基于建筑环境学课程的设计实践课程，意在将建筑环境学课程的理论知识通过设计案例的练习加以运用，加深对建筑环境学知识的理解，并训练如何通过设计优化营造良好室内环境品质。
课程分为授课和案例设计练习两部分，授课部分介绍目前关于被动式设计研究成果、工程实践案例中的被动式设计方法以及软件模拟分析技术；案例设计练习教授学生基于性能评估的优化设计方法，选取学生较为熟悉的住宅、幼儿园等体量较小的建筑类型作为设计优化对象，通过软件分析发现现有的室内环境设计不足，并基于现有研究成果提出优化策略，最后通过软件模拟加以验证。课程要求学生将建筑环境学课程所学知识用于本设计课程的室内环境品质的量化及控制。本课程着重训练建筑设计与环境工程学科知识的配合。

Course descriptions
This course is a design practice course based on the course of Building Environment, aiming to apply the theoretical knowledge of the Building Environment course through the practice of design cases, so as to deepen the understanding of the knowledge of building environment, and train how to create a good indoor environment quality through design optimization.
The course is divided into two parts: teaching and case design practice. The teaching part introduces current research results of passive design, passive design methods in engineering practice cases and software simulation analysis technology.Case design practice teaches students the optimal design method based on performance evaluation, choose the residence, kindergarten and other buildings with small volume that students are more familiar with as the design optimization objects, find existing deficiency in indoor environment design through software analysis, propose optimization strategies according to existing research results, and finally verify through software simulation. The course requires students to apply what they have learned in the Building Environment to the quantification and control of indoor environment quality in design.This course focuses on the integration of architectural design and environmental engineering knowledge.

研究生一年级
技术人文与建筑创新 • 窦平平
课程类型：选修
学时 / 学分：32 学时 / 2 学分

Graduate Program 1st Year
TECHNICAL HUMANITIES AND ARCHITECTURAL INNOVATION • DOU Pingping
Type: Elective Course
Study Period and Credits: 32 hours / 2 credits

教学内容
课程共 16 讲，分为 4 个主题系列和 1 项设计研究任务。每个系列有 3 个主题讲座组成，有学科前沿认识，有跨时代的纵向梳理，也有跨文化和跨学科的横向比较。设计研究任务在最后两周，由四次随堂讨论和课后自主组成。

教学方法
课程将采用国际一流大学的系列讲座、随堂研讨与设计研究实践相结合的教学方法。主题讲座包含系统知识要点、论点和案例分析，由任课教师主讲，注重内容的连贯性和专题性。课程将根据内容融入优质教学资源，为学生营造双语的教学环境。课堂播放课件采用结合图像、图表、影像、VR 等多种形式的数字化、信息化、可视化教学手段，使课堂讲授更加生动、清晰、深刻。同时注重网上教学资源的开发，根据情况灵活采用线上与线下相结合的教学模式。

Teaching contents
The course consists of 16 lectures, divided into four thematic series and one design research task. Each series consists of three themed lectures, with a cutting-edge understanding of the subject, a cross generational vertical review, and horizontal comparisons across cultures and disciplines. The design research task is completed in the last two weeks through four in class discussions and after class autonomy.

Teaching methods
The course will adopt a teaching method that combines a series of lectures, class discussions, and design research practices from world-class universities. The theme lecture is a systematic teaching of key knowledge points, arguments, and case studies, with a focus on the coherence and topicality of the content. The course will integrate high-quality teaching resources based on the content, creating a bilingual teaching environment for students. Classroom courseware adopts various forms of digital, informational, and visual teaching methods such as images, charts, videos, and VR, making classroom teaching more vivid, clear, and profound. At the same time, attention should be paid to the development of online teaching resources, and a flexible combination of online and offline teaching modes should be adopted according to the situation.

研究生一年级
建筑环境学 • 郜志
课程类型：选修
学时 / 学分：36 学时 / 2 学分

Graduate Program 1st Year
BUILDING ENVIRONMENT • GAO Zhi
Type: Elective Course
Study Period and Credits: 36 hours / 2 credits

教学目标
本课程的主要任务是使学生掌握建筑环境的基本概念，学习建筑热湿环境和空气环境的基础知识。通过课程教学，使学生熟悉建筑外环境、热湿环境的理论，并了解人体对热湿环境的反应，掌握建筑环境学的实际应用和最新研究进展，为建筑能源和环境系统的测量与模拟打下坚实的基础。

Teaching objectives
The main task of this course is to enable students to master the basic concepts of building environment and learn the basic knowledge of building thermal and humid environment and air environment. Through course teaching, students are familiarized with the theory of building external environment and thermal and humid environment, and understand the human body's response to thermal and humid environment. They master the practical application and latest research progress of building environmental science, and lay a solid foundation for the measurement and simulation of building energy and environmental systems.

本科三年级
乡村振兴建设实践 · 梁宇舒
课程类型：选修
学时 / 学分：32 学时 / 2 学分

Undergraduate Program 3rd Year
PRACTICE OF RURAL REVITALIZATION
CONSTRUCTION • LIANG Yushu
Type: Elective Course
Study Period and Credits: 32 hours / 2 credits

教学目标
　　本课程的任务主要是通过乡村现场调研以及驻场设计实践，培养学生借助多学科手段分析和认知实际社会问题的能力，并且综合运用建筑设计、乡村规划等方法，针对实际场景提出可行的解决方案，加深图纸设计与实际建造、社会服务之间的关联性理解。

课程内容
1. 行前培训讲座；
2. 乡村调研及民居考察；
3. 乡村调研报告讨论；
4. 乡村振兴规划与建筑设计；
5. 乡村振兴工作营成果汇报。

Teaching objectives
The main task of this course is to cultivate students' ability to analyze and recognize practical social problems through multi-disciplinary methods according to rural on-site research and on-site design practice, and to comprehensively apply methods such as architectural design and rural planning to propose feasible solutions for practical scenarios, deepening the understanding of the correlation between drawing design, actual construction, and social services.

Course contents
1. Pre-departure training lectures;
2. Rural research and residential investigation;
3. Discussion on rural research reports;
4. Rural revitalization planning and architectural design;
5. Reports on the achievements of the rural revitalization work camp.

本科四年级
工地实习 · 傅筱
课程类型：选修
学时 / 学分：36 学时 / 2 学分

Undergraduate Program 4th Year
PRACTICE IN CONSTRUCTION PLANT
• FU Xiao
Type: Elective Course
Study Period and Credits: 36 hours / 2 credits

教学目标
　　本课程的任务主要是通过工地现场考察，让学生对建筑实际建造流程有一定的直观认识；并要求对某一建筑构件的建造过程进行实地观察，然后绘制建筑构件施工图纸，加深图纸设计与实际建造之间的关联性理解。

课程内容
1. 建筑施工基本流程；
2. 建筑工地实习（一）——观察；
3. 建筑工地观察报告讨论；
4. 建筑工地实习（二）——构件测绘；
5. 建筑构件施工图纸设计。

课程要求
1. 讲授大纲的重点内容；
2. 通过实地考察，训练学生的观察分析能力，培养学生树立将理论知识向实践转化的主观意识；
3. 通过课程设计，加深学生对专业理论知识与工程实践相互关系的理解，提高其分析、解决实际问题的能力。

Instructional objectives
The main task of this course is to provide students with a certain intuitive understanding of the actual construction process of buildings through on-site inspection; And it requires on-site observation of the construction process of a certain building component, and then draw construction drawings of the building component to deepen the understanding of the correlation between drawing design and actual construction.

Course contents
1. Basic construction process;
2. Construction site internship (1)—Observation;
3. Discussion of construction site observation reports;
4. Construction site internship (II)—Component surveying and mapping;
5. Design of construction drawings for building components.

Course requirements
1. Teach key content of the teaching outline.
2. Through on-site inspection, train students' observation and analysis abilities, and cultivate their subjective awareness of transforming theoretical knowledge into practice.
3. Through course design, deepen students' understanding of the relationship between professional theoretical knowledge and engineering practice, and improve their ability to analyze and solve practical problems.

本科四年级
古建筑测绘 · 赵辰　史文娟
课程类型：选修
学时 / 学分：36 学时 / 2 学分

Undergraduate Program 4th Year
SURVEY AND DRAWING OF ANCIENT
BUILDINGS • ZHAO Chen　SHI Wenjuan
Type: Elective Course
Study Period and Credits: 36 hours / 2 credits

教学目标
　　本课程是建筑学专业本科生的专业基础理论课程。本课程的任务是使学生切实理解中国传统建筑结构体系、构造关系及比例尺度等基本概念，培养学生对传统建筑年代鉴定和价值判断的基本技能。

课程内容
　　通过室外作业和室内工作两个阶段，完成现场测绘和整理图纸报告两个环节。"测"，由现场实物的尺寸数据的观测量取，"绘"：根据测量数据与草图进行处理、整理绘制备的测绘图纸，最终完成个人独立的答辩报告。
阶段一：抵达现场、讲座
阶段二：现场工作——各小组分工协作
阶段三：图纸绘制——小组资料归档、排版
阶段四：答辩准备——打印图纸、个人独立研究
阶段五：交图答辩

Teaching objectives
This course is a fundamental theoretical course for undergraduate students majoring in architecture. The task of this course is to enable students to really understand the basic concepts of the structural system, structural relationship and scale of traditional Chinese buildings, and to develop students' basic skills of age identification and value judgment on traditional buildings.

Course contents
Through two stages of outdoor and indoor work, on-site surveying and mapping and drawing report organizing are completed: "measuring" means: taking measurements based on the size data of on-site physical objects; "drawing" means: processing and organizing complete surveying and mapping drawings based on measurement data and sketches. Ultimately complete an independent defense report.
Stage 1: Arrival at the site, lecture
Stage 2: On-site work—Dividing the tasks and collaboration among various groups
Stage 3: Drawing—Archiving and layout of group materials
Stage 4: Defense preparation—Print drawings and conduct individual independent research
Stage 5: Submission of drawings for defense

本科三年级
城乡认知实习 · 冯建喜 王洁琼
课程类型：选修
学时 / 学分：32 学时 / 2 学分

Undergraduate Program 3rd Year
URBAN AND RURAL COGNITIVE INTERNSHIP
• Feng Jianxi, WANG Jieqiong
Type: Elective Course
Study Period and Credits: 32 hours / 2 credits

教学目标
1. 系统的学习区域、城市与建筑认知的理念与方法，同时深入了解建筑设计与评价的基本方法和原则。
2. 了解典型城市区域关系、发展脉络、用地布局与空间结构、城市优秀地段和建筑空间的尺度和形式。
3. 学习成功的区域规划、城市和建筑设计的范例，也认知失败的区域关系、城市空间和建筑，从中汲取经验教训，增加感性认识，开阔眼界。
4. 掌握评价现实城市与建筑的方法，独立完成城市和建筑空间的认知分析报告及学术论文。
5. 熟悉具体区域规划、城市规划和建筑设计过程中解决实际问题的方法，同时为下一阶段的城市规划专业学习打下良好的基础。

Teaching objectives
1. Systematically study the concepts and methods of regional, urban, and architectural cognition, while diving into the basic methods and principles of architectural design and evaluation.
2. Understand typical urban regional relationships, development context, land layout and spatial structure, as well as the scale and form of excellent urban areas and architectural space.
3. Learn examples of successful regional planning, urban and architectural design, as well as recognize failed regional relationships, urban space and architecture, draw lessons from them, increase emotional understanding, and broaden horizons.
4. Master the methods of evaluating practical cities and buildings, independently complete cognitive analysis reports and academic papers on urban and architectural space.
5. Familiar with the methods of solving practical problems in specific regional planning, urban planning, and architectural design, while laying a solid foundation for the next stage of urban planning learning.

其他
MISCELLANEA

讲座
Lectures

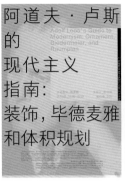
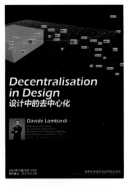

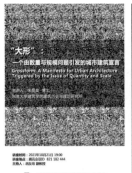

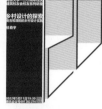
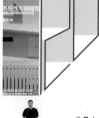

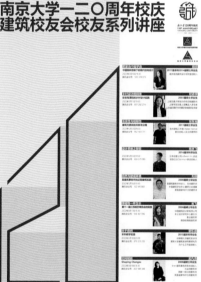

硕士学位论文列表
List of Thesis for Master Degree

研究生姓名	研究生论文标题	导师姓名
曹焱	城市住区形态性能指标研究	丁沃沃
况赫	基于视锥射线模型的街道空间形态特征量化研究	丁沃沃
吕文倩	容积率与街廓尺度、功能及路网形态的关联性研究	丁沃沃
王若辰	研发类建筑物理环境优化设计研究——以深时数字地球国际卓越研究中心为例	丁沃沃
辛宇	基于容积率的街廓物理环境性能评估——以南京为例	丁沃沃
卞真	南方山地历史乡镇与快速道路交通的形态重整研究	赵辰
皇甫子玥	关于水与南方山地历史聚落的形态关系研究	赵辰
莫默	浅析杨廷宝的"中国固有式"建筑"立面设计"——以金陵大学图书馆为例	赵辰
濮文睿	存量发展背景下的校城空间关系——以南京大学鼓楼校区为例	赵辰
郑经纬	"福安大厝"的类型研究	赵辰
周诗琪	基于人体模度的家具结构集成化的建筑构架体系研究	赵辰
谭锦楠	无梁殿——中国传统建筑语境中的砌体建筑	王骏阳
王旭	文丘里与塚本由晴的城市研究和建筑实践的关系的比较研究	王骏阳
陈一帆	基于深度学习的建筑形态分类识别方法研究	吉国华
匡鑫	基于体素的建筑形体生成研究	吉国华
李嘉伟	基于图论和拓扑关系限定的建筑群体体量布局生成方法研究	吉国华
罗逍遥	基于多目标优化的建筑外遮阳形式研究	吉国华
王家洲	基于室内光环境分析优化的开窗形式设计	吉国华
张含	体量与立面协同的建筑性能优化设计方法研究	吉国华
丁展图	后疫情时代厂区办公建筑空间设计研究——以现代汽车集团氢能源电池厂项目为例	张雷
刘贺	基本建筑在苏北农房设计中的应用研究——以宿迁秦庄新型农村社区项目为例	张雷
史鑫尧	主题书店的在地性设计研究——以李庄国际建筑书店和营造学社学术交流中心为例	张雷
谭路路	筒形工业遗产构筑物改造利用研究——以南京南部新城油库公园改造为例	张雷
王锴	中高密度用地条件下艺术中学公共空间设计——以广州艺术中学为例	张雷
王赛施	熙南里工作室改造及复合界面设计研究	张雷
陈志凡	南京大学鼓楼校区幼儿园改造设计——被动式节能设计研究	冯金龙
李乐	南京大学苏州校区地球系统与未来环境学科群共享平台设计——学科群建筑共享空间设计研究	冯金龙
廖伟平	南京大学苏州校区地球系统与未来环境学科群实验楼设计——地学实验室建筑模块化设计策略研究应用	冯金龙
朱维韬	南京市龙王山小学建筑设计——校医建筑群组织与气候适应性研究	冯金龙
朱晓晨	南京大学鼓楼校区树华楼知行楼改造更新设计——外立面陶砖幕墙构造研究	冯金龙
戈可辰	江苏瀚鸿电器现代化物流基地配套酒店设计——基于绿色酒店的被动式设计策略研究	冯金龙、谢明瑞
胡应航	江苏瀚鸿电器物流基地办公楼建筑幕墙设计策略研究	冯金龙、谢明瑞
车娟娟	基于形态类型学的浙江嘉善丁栅集镇肌理织补策略研究	周凌
何璇	地域文化视角下特色传统村落的规划设计策略研究——以盐城市收成老村规划设计为例	周凌
黄菲	新教育理念下中学建筑公共空间及教学单元探究——以南京市雨花台区某初中设计为例	周凌
黄晓寻	基于形态类型学方法的黔南荔波古镇老街历史肌理修复与公共空间设计	周凌
宋晓宇	南京金牛湖理想村规划与建筑设计	周凌

研究生姓名	研究生论文标题	导师姓名
张 尊	高校综合体设计研究——以南京大学苏州校区科研综合体设计为例	周 凌
岑国桢	南京市溧水区南极星产业园展示中心改扩建设计——大进深办公建筑光环境优化	傅 筱
陈玉珊	可拆装设计在建筑围护体系中的应用初探	傅 筱
董 青	小型集群式办公园区规划布局及单体套型形态研究——南极星科创园项目方案设计	傅 筱
傅婷婷	金属薄板在建筑外围护结构中的构造设计及其表达——以燕园路社区服务中心为例	傅 筱
王梦兰	基层社区服务中心公共空间人性化细部设计研究——兴卫村基层社区服务中心室内设计	傅 筱
翁 昕	老旧社区幼儿园改造设计研究——以南京小贝壳幼儿园为例	傅 筱
王维依	基于智能技术的场所互联及其在建筑中的应用初探	鲁安东
冯杨帆	城市空间实验室设计研究	鲁安东
谷雨阳	历史街区环境中带叙事的空间要素在AR中的应用研究——以南京颐和路历史街区为例	鲁安东
李博雅	城市历史人文空间数据库原型——一种设计辅助决策方法及其应用初探	鲁安东
王新强	1949年后垛田形态演变机理研究——以东旺村、张皮村为例	鲁安东
杨子媛	基于Mapping手段解读城市更新语境下日常公共空间——以南京市门西荷花塘居住型历史文化街区为例	鲁安东
洪 静	坦比哀多小圣堂与圣彼得大教堂——论文艺复兴时期集中式教堂观念的成型	胡 恒
李伊萌	佛罗伦萨主教堂穹顶采光亭研究	胡 恒
范嫣琳	苏南70—80年代狭长型联排农居更新设计研究——常州市横林镇狄坂村农居改造设计	华晓宁
侯自忠	不同发育类型的圩田景观中村落形态比较研究——以杭嘉湖东部圩区为例	华晓宁
李心仪	西藏林芝地区建筑地域性的现代表达——林芝索松村度假酒店方案设计	华晓宁
臧 哲	乡村农产品加工综合园区空间组织模式研究——马鞍山博望区特色农业产业园方案设计	华晓宁
明文静	基于太阳能和天然采光综合利用的光伏天窗应用研究——以南京旧工业建筑改造为例	吴 蔚
程 绪	基于Cesium的老旧小区轻量级CIM搭建研究——以南京饮虹园片区为例	童滋雨
李 昂	基于三阶段DEA模型的环境因素对街道步行空间使用效率的影响研究	童滋雨
李昌曦	面向老旧小区更新的CIM数据库框架构建研究	童滋雨
沈育辉	基于大数据的人本尺度社区多层级生活圈便利性评估	童滋雨
陈星雨	内容计划（program）设计方法的研究	胡友培
李 雪	南京市饮虹园历史风貌地段居住改善类城市更新设计研究	胡友培
王子涵	基于街巷空间活化的历史风貌地段城市更新设计研究——以南京饮虹园为例	胡友培
魏雪仪	奥斯瓦尔德·马蒂亚斯·昂格尔斯城市设计思想与方法的研究	胡友培
程 蕙	形态学视野下文化与民生双驱动的城市更新单元划分研究——以苏州古城双塔片区为例	窦平平
戴添趣	近郊产业融合下的新型空间复合模式探索——以南京八卦洲某全产业链示范园为例	窦平平
朱 硕	面向疗愈环境的建筑与景观综合设计研究	窦平平
孔 严	中央体育场旧址田径场建设背景及意义初探	冷 天
李芸梦	基于虚拟复原的历史信息整合研究——以"汇文书院"校门及体育馆为例	冷 天
袁 琴	金陵大学近代西式住宅空间要素及其组织方式初探	冷 天
陈莉莉	典型被动式节能技术在江苏夏热冬冷地区乡村住宅中的适用性研究——以句容陈庄村为例	梁卫辉
温 琳	建筑"水-绿"空间的整合设计研究——以南京六合姚庄片区社区中心方案设计为例	刘 铨、冯金龙
梁 颖	南京河流沿岸聚落肌理的形态特征研究	刘 铨、丁沃沃

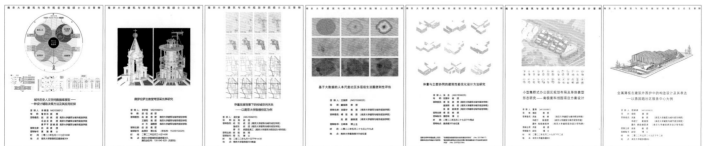

在校学生名单
List of Students

本科生 Undergraduate

2018级学生 / Students 2018

包诗贤 BAO Shixian	林济武 LIN Jiwu	邱雨欣 QIU Yuxin	肖郁伟 XIAO Yuwei	张 同 ZHANG Tong
陈锐娇 CHEN Ruijiao	刘瑞翔 LIU Ruixiang	沈 洁 SHEN Jie	熊浩宁 XIONG Haoyu	张新雨 ZHANG Xinyu
冯德庆 FENG Deqing	刘湘菲 LIU Xiangfei	宋佳艺 SONG Jiayi	徐 颖 XU Ying	周宇阳 ZHOU Yuyang
顾嵘健 GU Rongjian	陆麒竹 LU Qizhu	孙穆群 SUN Muqun	薛云龙 XUE Yunlong	阿尔申·巴特尔江 Aershen BATEERJIANG
顾祥姝 GU Xiangshu	罗宇豪 LUO Yuhao	田 靖 TIAN Jing	杨 朵 YANG Duo	
何 旭 HE Xu	倪梦琪 NI Mengqi	田舒琳 TIAN Shulin	喻姝凡 YU Shufan	
李逸凡 LI Yifan	牛乐乐 NIU Lele	吴高鑫 WU Gaoxin	张百慧 ZHANG Baihui	

2019级学生 / Students 2019

高禾雨 GAO Heyu	石珂千 SHI Keqian	周昌赫 ZHOU Changhe
高赵龙 GAO Zhaolong	唐诗诗 TANG Shishi	上原舜平 SHANGYUAN Shunping
顾 靓 GU Liang	王思戎 WANG Sirong	麦吾兰江·穆合塔尔 Maiwulanjiang MUHETER
黄辰逸 HUANG Chenyi	王智坚 WANG Zhijian	
黄小东 HUANG Xiaodong	王梓蔚 WANG Ziwei	
黄煜东 HUANG Yudong	袁 泽 YUAN Ze	
邱雨婷 QIU Yuting	张楚杭 ZHANG Chuhang	

2020级学生 / Students 2020

陈浏毓 CHEN Liuyu	华羽纶 HUA Yuguan	刘卓然 LIU Zhuoran	吴嘉文 WU Jiawen
陈沈婷 CHEN Shenting	黄淑睿 HUANG Shurui	陆星宇 LU Xingyu	袁欣鹏 YUAN Xinpeng
陈 玚 CHEN Cheng	李静怡 LI Jingyi	钱梦南 QIAN Mengnan	杨曦睿 YANG Xirui
陈璇霖 CHEN Xuanlin	李沛熹 LI Peixi	沈至文 SHEN Zhiwen	张嘉木 ZHANG Jiamu
高 晴 GAO Qing	李若松 LI Ruosong	孙昊天 SUN Haotian	张伊儿 ZHAGN Yi'er
顾 林 GU Lin	刘珩歆 LIU Hengxin	王天赐 WANG Tianci	
何德林 HE Delin	刘晓斌 LIU Xiaobin	王天歌 WANG Tiange	

2021级学生 / Students 2021

阿 旦 A Dan	黄静雯 HUANG Jingwen	李梦麟 LI Menglin	阮锦涛 RUAN Jintao	王翔宇 WANG Xiangyu	杨子江 YANG Zijiang	周锦润 ZHOU Jinrun
步欣洁 BU Xinjie	黄思宇 HUANG Siyu	李宇博 LI Yubo	佘柯宇 SHE Keyu	王艺楠 WANG Yinan	张纪闻 ZHANG Jiwen	周抒秋 ZHOU Shuqiu
陈 郝 CHEN He	黄文郡 HUANG Wenjun	黎宇航 LI Yuhang	申 翱 SHEN Ao	王雨晴 WANG Yuqing	张芮宁 ZHANG Ruining	朱 轶 ZHU Yi
陈 可 CHEN Ke	江炫烨 JIANG Xuanye	李雨芊 LI Yuqian	唐扬航 TANG Yanghang	奚 凡 XI Fan	张昱程 ZHANG Yucheng	邹子午 ZOU Ziwu
陈雅琪 CHEN Yaqi	金 毅 JIN Yi	刘新桐 LIU Xintong	万飞扬 WAN Feiyang	夏知愚 XIA Zhiyu	赵 通 ZHAO Tong	
陈宇轩 CHEN Yuxuan	孔令璇 KONG Lingxuan	刘子宣 LIU Zixuan	万家良 WAN Jialiang	颜星林 YAN Xinglin	郑可薇 ZHENG Kewei	
何若然 HE Ruoran	李俊杰 LI Junjie	彭 鑫 PENG Xin	汪祉健 WANG Zhijian	杨雯文 YANG Wenwen	周皓誉 ZHOU Haoyu	

研究生 Postgraduate

卞真 BIAN Zhen	程绪 CHENG Xu	戈可辰 GE Kechen	黄晓寻 HUANG Xiaoxun	李嘉伟 LI Jiawei	刘贺 LIU He	宋晓宇 SONG Xiaoyu	王若辰 Wang Ruochen	辛宇 XIN Yu	周诗琪 ZHOU Shiqi
岑国桢 CEN Guozhen	程懿 CHENG Yi	谷雨阳 GU Yuyang	皇甫子玥 HUANGFU Ziyue	李心仪 LI Xinyi	吕文倩 Lü Wenqian	孙其 SUN Qi	王赛施 WANG Saishi	杨子媛 YANG Ziyuan	朱硕 ZHU Shuo
车娟娟 CHE Juanjuan	戴添趣 DAI Tianqu	顾梦婕 GU Mengjie	孔严 KONG Yan	李雪 LI Xue	罗逍遥 LUO Xiaoyao	谭锦楠 TAN Jinnan	王新强 WANG Xinqiang	袁琴 YUAN Qin	朱维韬 ZHU Weitao
陈莉莉 CHEN Lili	丁展图 DING Zhantu	何璇 HE Xuan	匡鑫 KUANG Xin	李伊萌 LI Yimeng	明文静 MING Wenjing	谭路路 TAN Lulu	王旭 WANG Xu	臧哲 ZANG Zhe	朱晓晨 ZHU Xiaochen
陈星雨 CHEN Xingyu	董青 DONG Qing	洪静 HONG Jing	况赫 KUANG He	李乐 LI Yue	莫默 MO Mo	王家洲 WANG Jiazhou	王子涵 WANG Zihan	张含 ZHANG Han	
陈一帆 CHEN Yifan	范嫣琳 FANG Yanlin	侯自忠 HOU Zizhong	李昂 LI Ang	李芸梦 LI Yunmeng	濮文睿 PU Wenrui	王锴 WANG Kai	温琳 WEN Lin	张尊 ZHANG Zun	
陈玉珊 CHEN Yushan	冯杨帆 FENG Yangfan	胡应航 HU Yinghang	李博雅 LI Boya	梁颖 LIANG Yin	沈育辉 SHEN Yuhui	王梦兰 WANG Menglan	翁昕 WENG Xin	赵彤 ZHAO Tong	
陈志凡 CHEN Zhifan	傅婷婷 FU Tingting	黄菲 HUANG Fei	李昌曦 LI Changxi	廖伟平 LIAO Weiping	史鑫尧 SHI Xinyao	王鹏程 WANG Pengcheng	魏雪仪 WEI Xueyi	郑经纬 ZHENG Jingwei	

白珂嘉 BAI Kejia	丁嘉欣 DING Jiaxin	蒋哲 JIANG Zhe	刘奕孜 LIU Yizi	丘雨辰 QIU Yuchen	王路 WANG Lu	邢雨辰 XING Yuchen	余沁蔓 YU Qinman	赵琳芝 ZHAO Linzhi	
陈婧秋 CHEN Jingqiu	丁明昊 DING Minghao	赖泽贤 LAI Zexian	刘雨田 LIU Yutian	仇佳豪 QIU Jiahao	王明珠 WANG Mingzhu	徐福锁 XU Fusuo	于文爽 YU Wenshuang	赵亚迪 ZHAO Yadi	
陈铭行 CHEN Mingxing	方奕璇 FANG Yixuan	雷畅 LEI Chang	罗紫娟 LUO Zijuan	任钰佳 REN Yujia	王琪 WANG Qi	徐佳楠 XU Jianan	于智超 YU Zhichao	赵子文 ZHAO Ziwen	
陈茜 CHEN Qian	费元丽 FEI Yuanli	李昂 LI Ang	吕广彤 Lü Guangtong	邵桐 SHAO Tong	王瑞蓬 WANG Ruipeng	许龄 XU Ling	袁振香 YUAN Zhenxiang	周理洁 ZHOU Lijie	
陈颖 CHEN Ying	冯智 FENG Zhi	李倩 LI Qian	马丹艺 MA Danyi	盛泽明 SHENG Zeming	王雨嘉 WANG Yujia	杨东来 YANG Donglai	曾敬淇 ZENG Jingqi	周宇飞 ZHOU Yufei	
陈宇帆 CHEN Yufan	龚豪辉 GONG Haohui	李晓云 LI Xiaoyun	马致远 MA Zhiyuan	宋贻泽 SONG Yize	翁鸿炜 WENG Hongyi	杨岚 YANG Lan	张梦冉 ZHANG Mengran	朱辰浩 ZHU Chenhao	
陈予婧 CHEN Yujing	龚泰冉 GONG Tairan	林之茜 LIN Zhiqian	潘晴 PAN Qing	孙杰 SUN Jie	吴子豪 WU Zihao	杨瑞凯 YANG Ruikan	张塑琪 ZHANG Suqi	朱激清 ZHU Jiqing	
程慧 CHENG Hui	胡永裕 HU Yongyu	刘亲贤 LIU Qinxian	庞馨怡 PANG Xinyi	唐敏 TANG Min	谢菲 XIE Fei	杨淑铌 YANG Shuchan	张云松 ZHANG Yunsong	朱凌云 ZHU Lingyun	
崔晓伟 CUI Xiaowei	黄翊健 HUANG Yijie	刘雪寒 LIU Xuehan	秦钢强 QIN Gangqiang	王蕾 WANG Lei	谢文俊 XIE Wenjun	于瀚清 YU Hanqing	赵济宁 ZHAO Jining	朱孟阳 ZHU Mengyang	

陈露茜 CHEN Luxi	傅峻岩 FU Junyan	李静娴 LI Jingxian	白雪 BAI Xue	郭珊 GUO Shan	龙千慧 LONG Qianhui	邱国强 QIU Guoqiang	王春磊 WANG Chunlei	魏高翔 WEI Gaoxiang	杨佳锟 YANG Jiakun	章超 ZHANG Chao	钟子超 ZHONG Zichao
陈宜旻 CHEN Yimin	高东旸 GAO Dongyang	李静愉 LI Jingyu	曹超 CAO Chao	蒋东梁 JIANG Donglin	卢禹成 LU Zhuocheng	邱向楠 QIU Xiangnan	王鲁 WANG Lu	谢宇航 XIE Yuhang	杨茸佳 YANG Rongjia	张璐 ZHANG Lu	周航 ZHOU Hang
程科懿 CHENG Keyi	顾渫非 GU Xiefei	李鹿 LI Lu	陈威霖 CHEN Weilin	孔捷 KONG Jie	骆婧雯 LUO Jingwen	仇凯莹 QIU Kaiying	王琪泓 WANG Qihong	徐小越 XU Xiaoyue	杨晟铨 YANG Shengquan	张鹏 ZHANG Peng	周金雨 ZHOU Jinyu
董冰涛 DONG Bingtao	郭浩哲 GUO Haozhe	李帅 LI Shuai	陈卓 CHEN Zhuo	李瑾 LI Jin	罗婷 LUO Ting	邵鑫露 SHAO Xinlu	王瑞明 WANG Ruiming	许雁庭 XU Yanting	杨乙彬 YANG Yibin	赵茂繁 ZHAO Maofan	周盟珊 ZHOU Mengshan
范玉斌 FAN Yubin	郝昕苑 HAO Xinyuan	李伟 LI Wei	崔伊瑄 CUI Yixuan	廖艺敏 LIAO Yimin	马骧 MA Ziang	孙弘睿 SUN Hongrui	王潇 WANG Xiao	许一凡 XU Yifan	杨子征 YANG Zizheng	赵文 ZHAO Wen	祝诗雅 ZHU Shiya
范悦 FAN Yue	和煦 HE Xu	李文秀 LI Wenxiu	邓诺怡 DENG Nuoyi	林文倬 LIN Wenzhuo	麦思琪 MAI Siqi	孙珂 SUN Ke	王晓茜 WANG Xiaoqian	闫朝新 YAN Chaoxin	姚孟君 YAO Mengjun	赵晓雪 ZHAO Xiaoxue	朱颜怡 ZHU Yanyi
方雨 FANG Yu	胡峻语 HU Junyu	李宇翔 LI Yuxiang	邓泽旭 DENG Zexu	刘鑫睿 LIU Xinrui	彭昊韦 PENG Haowei	孙志伟 SUN Zhiwei	王奕绮 WANG Yiqi	晏攀 YAN Pan	叶庆锋 YE Qingfeng	赵越 ZHAO Yue	
冯子恺 FENG Zikai	黄柯 HUANG Ke	林晴 LIN Qing	董志昀 DONG Zhiyun	刘玥蓉 LIU Yuerong	彭洋 PENG Yang	王奔 WANG Ben	王玥 WANG Yue	杨帆 YANG Fan	翟罂钰 ZHAI Zhaoyu	钟言 ZHONG Yan	

图书在版编目（CIP）数据

南京大学建筑与城市规划学院建筑系教学年鉴：2021—2022 / 唐莲，李鑫编. -- 南京：东南大学出版社，2024.6
ISBN 978-7-5766-1139-7

Ⅰ.①南… Ⅱ.①唐…②李… Ⅲ.①南京大学-建筑学-教学研究-2021-2022-年鉴 Ⅴ.①TU-42

中国国家版本馆CIP数据核字(2024)第014510号

编　委　会：丁沃沃　赵　辰　吉国华　周　凌　鲁安东　华晓宁　唐　莲　李　鑫
版面制作：唐　莲　李　鑫　赵茂繁　夏　月　黄辰逸　邱雨婷　刘佳慧
责任编辑：姜　来　魏晓平
责任校对：张万莹
封面设计：王丹丹　夏　月
责任印制：周荣虎

南京大学建筑与城市规划学院建筑系教学年鉴 2021—2022
Nanjing Daxue Jianzhu Yu Chengshi Guihua Xueyuan Jianzhuxi Jiaoxue Nianjian 2021-2022

出版发行：东南大学出版社
出 版 人：白云飞
社　　址：南京市四牌楼2号
网　　址：http://www.seupress.com
邮　　箱：press@seupress.com
邮　　编：210096
电　　话：025-83793330
经　　销：全国各地新华书店
印　　刷：南京新世纪联盟印务有限公司
开　　本：889 mm×1 194 mm　1/20
印　　张：11
字　　数：436千
版　　次：2024年6月第1版
印　　次：2024年6月第1次印刷
书　　号：ISBN 978-7-5766-1139-7
定　　价：78.00元

本社图书若有印装质量问题，请直接与营销部联系。电话：025-83791830